Defence and Discovery

Studies in Canadian Military History
Series editor: Dean F. Oliver, Canadian War Museum

The Canadian War Museum, Canada's national museum of military history, has a threefold mandate: to remember, to preserve, and to educate. Studies in Canadian Military History, published by UBC Press in association with the Museum, extends this mandate by presenting the best of contemporary scholarship to provide new insights into all aspects of Canadian military history, from earliest times to recent events. The work of a new generation of scholars is especially encouraged, and the books employ a variety of approaches – cultural, social, intellectual, economic, political, and comparative – to investigate gaps in the existing historiography. The books in the series feed immediately into future exhibitions, programs, and outreach efforts by the Canadian War Museum. A list of the titles in the series appears at the end of the book.

Defence and Discovery
Canada's Military Space Program, 1945-74

Andrew B. Godefroy

UBCPress · Vancouver · Toronto

20 19 18 17 16 15 14 13 12 11   5 4 3 2 1

Printed in Canada on FSC-certified ancient-forest-free paper (100% post-consumer recycled) that is processed chlorine- and acid-free.

Library and Archives Canada Cataloguing in Publication

Godefroy, Andrew B., 1972-
Defence and discovery : Canada's military space program, 1945-74 / Andrew B. Godefroy.

(Studies in Canadian military history, ISSN 1499-6251)
Includes bibliographical references and index.
ISBN 978-0-7748-1959-6 (bound); 978-0-7748-1960-2 (pbk)

1. Astronautics, Military – Canada – History. 2. Outer space – Exploration – Canada – History. 3. Canada – History, Military – 20th century. I. Title. II. Series: Studies in Canadian military history

UG1525.C3G63 2011   623'.690971   C2011-900314-7

e-book ISBNs: 978-0-7748-1961-9 (pdf); 978-0-7748-1962-6 (e-pub)

Canadä

UBC Press gratefully acknowledges the financial support for our publishing program of the Government of Canada (through the Canada Book Fund), the Canada Council for the Arts, and the British Columbia Arts Council.

This book has been published with the help of a grant from the Canadian Federation for the Humanities and Social Sciences, through the Aid to Scholarly Publications Program, using funds provided by the Social Sciences and Humanities Research Council of Canada.

Publication of this book has been financially supported by the Canadian War Museum.

Printed and bound in Canada by Friesens
Set in Helvetica Condensed and Minion by Artegraphica Design Co. Ltd.
Copy editor: Deborah Kerr
Proofreader: Lesley Erickson

UBC Press
The University of British Columbia
2029 West Mall
Vancouver, BC V6T 1Z2
www.ubcpress.ca

Contents

Figures

Acknowledgments

As is so often the case with research projects that eventually become books, the largest contributor to the success of this effort was passion. For as long as I can remember, I have looked into the night sky with a sense of bewilderment and pure wonder. More importantly, perhaps, as a child I relished the arrival of the annual family vacation to Florida, where at least one day of our retreat to Daytona was devoted to taking a very excited little boy down the coast and across the Indian River lagoon to the Kennedy Space Center. With a freshly stamped admission ticket in hand, I was unleashed among the exhibits to marvel at satellites and space capsules, lunar buggies and moon rocks. The exhibition on the "space transportation system" was my favourite. When I first observed this technological marvel in 1979, the space shuttle as it became commonly known had not yet launched into orbit. But the conceptual artwork surrounding the three-metre-tall model of the shuttle piggybacked on its boosters assured me that, when I was a "grown-up," I had a very good chance of living and working in outer space just like Dr. Heywood R. Floyd in the famous movie *2001: A Space Odyssey*.

Despite still being grounded later in life, I continued to ask myself "How did we get into space?" and "Who or what might we some day discover there?" I knew, perhaps with a hint of disappointment, that the latter question would probably not be answered in my lifetime, but as a student of history, I did know that the former question could be more easily tackled. My undergraduate studies revealed to me that the origins of rocketry and spaceflight were born out of matters of defence, and upon entering the War Studies Program at the Royal Military College of Canada, Kingston, Ontario, in 1997, I decided to focus my graduate and post-graduate studies on science, technology, and international relations, with a particular interest in missile, rocketry, and space programs.

This book had its genesis in my earlier examinations of the subject that focused upon much broader analyses of the relationship between the United States and Canada in matters related to missile and defence space programs. Among other things, it demonstrated that the two countries shared a rich history in scientific and technological cooperation, and that Canada's own efforts in early post-war rocketry, space science, and technology development were far more comprehensive than one might have thought. Though I spent years researching and writing this subject, in some ways I feel I have only scratched the surface. There is much more work to be done in this field, and what I have initiated here I hope other scholars will pursue even further.

I must thank many people for their professional encouragement throughout this journey, the most important being my mentor, adviser, and personal friend Joel Sokolsky. A leading authority on Canada-US relations, Joel freely shared his wealth of knowledge and worldly experience with me and provided an insight on the *realpolitik* of Canadian international relations that few others could. Joel encouraged my choice of subject, acknowledged the expected difficulties that lay ahead (such as material declassification), and nudged me to stay the course when other events invaded my thoughts and work. He showed great understanding when my studies were abruptly interrupted on 11 September 2001, as I was called away to active duty with the Canadian Forces Joint Operations Group, and he ensured that I was back on track the following year despite the ever-increasing demands of the Department of National Defence. I learned more from this gentleman than I could ever have hoped for and am forever grateful to Joel, a true scholar, who challenged me and encouraged me to seek answers, to explain and not to advocate, and most importantly to understand the Canadian way.

I owe much to the tremendous academic experience I gained from the War Studies Program at the Royal Military College of Canada and am especially grateful to Brian McKercher, then serving as war studies chair, for making it a "full-contact sport." Additionally, Jim Finan, Ron Haycock, Michael Hennessy, Hal Klepak, Sean Maloney, and Scot Robertson all provided me with food for thought throughout my course work and ensured that I was never left idle by giving me a healthy dose of studies to complete. Though I hated the extra workload at times, I realize now that it prepared me well for real world analysis, writing, and publishing in the defence sphere, where time is never a luxury, but there is no excuse for mediocre quality.

This study would not have been possible without the generous support of the Department of National Defence Security and Defence Forum and the War Studies Program, both of which provided assistance for my travel and field research. In the many places I visited to uncover the mysteries behind this history, I never encountered an inconsiderate or unhelpful response to my requests. Both NORAD and NATO offices worldwide catered to my inquiries for information as well as personal visits to the places where the history in this book unfolded. Library and Archives Canada lent considerable assistance to my research; I particularly thank those who handled my ceaseless requests for declassification of materials. The archives of the Department of Foreign Affairs and International Trade, the National Research Council, and the Directorate of History and Heritage were all very helpful. As well, I wish to acknowledge Magellan Aerospace, the current caretaker of Bristol's Black Brant legacy, for responding kindly to my requests for details of the long-lost rocket pioneers who developed Canada's early space launch technologies.

No study is ever finalized without the commentary and advice of peer review. I was especially fortunate to have Michael J. Neufeld, chair of the Space History Division at the Smithsonian's National Air and Space Museum and most recently the biographer of the legendary Wernher von Braun, who provided invaluable assistance in the genesis of this work. In the absence of any evolved space history expertise in Canada, Michael stepped away from a busy schedule at the centre of the American space history community to come to Kingston. I am very grateful to him for visiting and allowing me a chance to gain "street credit" in other academic circles.

Finally, I must thank UBC Press and in particular Emily Andrew for showing interest in this work from the beginning and supporting it through the publication process. The defence studies community in Canada remains isolated and underappreciated to some degree, but UBC Press has seized upon the opportunity to publish first-rate research from leading military historians in its military series. For that, I owe it my personal thanks. Any remaining errors in fact or opinion in this book are mine alone.

Abbreviations

ACSR	Associate Committee on Space Research
APT	automatic picture transmission
ARPA	Advanced Research Projects Agency
BAL	Bristol Aircraft (Western)
CARDE	Canadian Armament Research and Development Establishment
CAScW	Canadian Association of Scientific Workers
CEPE	Central Experimental and Proving Establishment
CFHQ	Canadian Forces Headquarters
CMSG	Canadian Military Space Group
COSINE	Co-Orbital Satellite Intercept Evaluation
CRPL	Central Radio Propagation Laboratory
CRPP	Canadian Rocket Propulsion Program
CRR	Churchill Research Range
CTS	chief of technical services
DAED	Directorate of Advanced Engineering and Development
DARPG	Development and Associated Research Policy Group
DDP	Department of Defence Production
DDR and E	Directorate of Defence Research and Engineering
DND	Department of National Defence
DOC	Department of Communications
DOD	Department of Defense
DRB	Defence Research Board
DRNL	Defence Research Northern Laboratory
DRTE	Defence Research Telecommunications Establishment
GCCSS	global commercial communications satellite system
ICBM	intercontinental ballistic missile
ICS	Interdepartmental Committee on Space
ICSC	Interim Canadian Space Council

IGY	International Geophysical Year
IPY	International Polar Year
ISIS	international satellites for ionospheric studies
LAC	Library and Archives Canada
MOSST	Ministry of State for Science and Technology
NASA	National Aeronautics and Space Administration
NATO	North Atlantic Treaty Organization
NORAD	North American Air Defence
NRC	National Research Council
PCO	Privy Council Office
PTV	Propulsion Test Vehicle
RCAF	Royal Canadian Air Force
RCN	Royal Canadian Navy
RPL	Radio Physics Laboratory
SIP	Space Indoctrination Program
SITU	Satellite Identification Tracking Unit
SPADATS	space detection and tracking system
STEM	storable tubular extendable member
TSWG	Topside Sounder Working Group
USAF	United States Air Force
USPACECOM	United States Space Command

Chronology of Canada's Rocketry and Space Program

c. 1916	National Research Council (NRC) created.
1940	National Resources Mobilization Act congregates most scientists in NRC to directly support Canada's war effort.
September 1945	Soviet cipher clerk and spy Igor Gouzenko defects to the Canadian government.
March 1947	Defence Research Board (DRB) created.
1949	NATO created.
February 1951	Defence Research Telecommunications Establishment (DRTE) created.
1956	CARDE reviews its existing propulsion testing and creates the Canadian Rocket Propulsion Program (CRPP).
1956-57	Canadian defence and civilian scientists participate in the International Geophysical Year (IGY)
October 1957	The USSR launches Sputnik, its first satellite.
1958	NORAD created. Albert Fia joins Bristol Aerospace and creates the Special Projects Group.
March 1959	Canada and the US cooperate to create the space detection and tracking system (SPADATS).
April 1959	Associate Committee on Space Research formed.
October 1959	First Propulsion Test Vehicle (PTV)/Black Brant I launched from the Churchill Research Range.
1960-61	RCAF Advanced Technology Evaluation Program created.
1961	Satellite Identification Tracking Unit (SITU) stood up at Cold Lake, Alberta.
1961	RCAF Space Indoctrination Program created.
1961	Canadian defence involvement in SPACETRACK begins.
1961	First launch of Black Brant IIA.
January 1961	Development of Black Brant III begins.
March 1961	Canadian Bristol Aerojet Ltd. formed.

June 1962	First launch of Black Brant III takes place at Wallops Island, Virginia.
September 1962	Canada's first satellite, Alouette I, is launched from Vandenberg AFB, California.
October 1962	Cuban Missile Crisis.
1963	Nuclear Test Ban Treaty signed.
1963	The Co-Orbital Satellite Intercept Evaluation (COSINE) Project begins.
1964	The RCAF Space Defence Program is officially initiated.
1964	Canadian Forces integration initiated.
April 1964	First Black Brant VA launched at the Churchill Research Range.
June 1964	First Black Brant IVA launched at the Churchill Research Range.
c. 1964-66	Science Secretariat created to advise government on space programs.
c. 1964-69	Alouette-ISIS project.
c. 1966	Science Council of Canada created.
1966	Science Secretariat's National Space Study.
April 1967	DRB/Bristol Aerojet Black Brant Development Program ends.
1967	Chapman Report issued.
1968	Department of National Defence unification initiated.
1968	Defence Research Telecommunications Establishment dissolved and re-created as Communications Research Center.
March 1968	*White Paper on a Domestic Satellite Communications System for Canada* published.
April 1968	Department of Communications (DOC) created.
January 1969	ISIS-A launched.
July 1969	Apollo XI moon landing.
January 1970	Interdepartmental Committee on Space first meeting.
April 1971	ISIS-B launched.
October 1971	Ministry of State for Science and Technology created.
November 1972	Anik A1 launched.
April 1973	Anik A2 launched.
1974	Canada ratifies its first national space policy.

Defence and Discovery

Introduction

THE DREAM OF REACHING outer space has teased the imaginations of humankind since the days of classic civilization, but only during the twentieth century did the knowledge and technologies needed to turn these dreams into fact became available. Yet, like so many great technological innovations of the last century, rocketry was brought to full fruition in total war and space flight during the political and strategic uncertainty that followed it.

Perhaps no other technological competition was a more popular icon of the early Cold War period than the "space race" between the United States and the Union of Soviet Socialist Republics. Fuelled as much by politics, fear, propaganda, and prestige as the need for strategic capability in surveillance, reconnaissance, and if necessary, military strike, the two superpowers engaged in a progressively challenging and expensive contest of innovation above the earth that changed the very nature of the Cold War itself and ultimately defined a new era in human civilization.[1]

Like a handful of other nations, Canada was an active participant in the early Cold War space race. Shortly after the Second World War, for reasons of self-preservation and self-interest, the country took its first steps toward the development of a national rocket and space program that, although not large, had tremendous vision and a high degree of technological success. Born out of scientific curiosity and shaped by technological innovation and the security requirements of the Cold War, the program in later decades fundamentally transformed Canada itself. Most importantly, however, Canada was able to parlay its own national technological competence in this field into political, military, and strategic saliency among its larger and more powerful allies, allowing it to play a considerable part during what was arguably the greatest epoch of discovery since the Renaissance and certainly the most dangerous period of human defence.

This book describes and analyzes Canada's role in the exploration and exploitation of the upper atmosphere and outer space between the end of the Second World War in 1945 and the ratification of the country's first national space policy in 1974. In doing so, it reveals the rich history of scientific and technological innovation in Canada during the early Cold War period and illuminates the central role of military enterprise in shaping dynamic technological change in this field during the post-war economic boom. Supporting other more recent interpretations of Canada's role as an active and astute participant within the Cold War Western alliance, this book assesses what Canada's strategic interests in outer space were within the wider context of the Cold War space race. Its findings demonstrate Canada's attempt at military self-reliance in the 1950s and its desire to remain a salient technical partner within its strategic alliances throughout the 1960s and early 1970s.

With the existence of an intimate relationship between its scientific, technological, defence, and government communities similar to those witnessed in the United States and Britain during this period, Canada attempted with some success to create a national rocketry and space program that would keep it relatively close to the leading edge of technological change as well as technologically relevant within the eyes of its primary strategic allies. Though technological achievement in the upper atmosphere and space was never the largest priority of the government during the early Cold War, Canada nevertheless took pride in it and applied its benefits wherever possible.

Until very recently, the portrayal of Canadian innovation in the post–Second World War period has made no substantial acknowledgment of the country's scientific and technological achievements. Until the late 1960s, Canadian political and economic historians generally characterized the country's evolution through a series of natural resource exploitations designed for global export that would later provide the basis for the creation of a mature manufacturing-based economy. The 1970s, however, saw a considerable shift in the interpretation of Canada's development. During that time, numerous government studies warned of a national failure to exploit industrialization, manufacturing, and trade, and a new generation of historians argued that Canada had failed to break away from its traditional dependence on raw material export after the Second World War. In fact, the country remained largely a semi-industrialized nation stuck in a pattern of arrested technological and economic development.

The catalyst of this new school of thought appeared in 1967 when historian J.J. Brown published *Ideas in Exile: A History of Canadian Invention*, an unflattering and at times damning history of Canadian innovative creation and technology. Though he commended Canadians individually for their inventiveness, he condemned the country as a whole for failing to capitalize on those inventions. Brown concluded that the lack of national innovation was a "psychological rather than financial problem" and described it as a fear of modernization and change. Finally, he castigated the government for neglecting to remedy the situation by encouraging and supporting innovation, as well as sustaining it through some form of succession planning.[2]

Why Brown suggested that Canadians failed to innovate after the Second World War is understandable from a certain perspective. Undoubtedly, the period during which he produced his work influenced him to some degree. In 1967 Canada was already embroiled in what would soon become a bitter debate over the future of its national science and space policies, and the post-war economic and technological expansion that had achieved much was beginning to plateau. At the same time, innovation was being subjected to cautious optimism, as both public and private organizations and investors awaited the outcome of federal policy decisions on the future of various national research and development objectives.

Yet, in the absence of widely published contrary views or arguments, Brown's thesis influenced both public opinion and the academic interpretation of the history of Canadian innovation for nearly fifteen years afterward. In 1982, however, economist Christian DeBresson successfully challenged the Brown thesis by re-examining Brown's data and revealing that a number of his conclusions were erroneous.[3] In particular, the existing evidence did not support Brown's claim that Canada had failed to innovate during the early Cold War period, leading DeBresson to suggest that a revision of the standard historical interpretation was desperately needed. Still, substantial case studies of post-war technological change in Canada remained sparse during the 1980s, and DeBresson was unable to develop his work on the subject. It was not perhaps until the mid-1990s, with the declassification of archives, that scholarly research became more feasible.

Scholars have increasingly argued that military establishments in industrialized nations have played an important historical role in shaping innovation as well as technological and economic change by linking national defence with national welfare. American historians Merritt Roe Smith and Alex Roland have both underscored the convergence of military and economic forces in US Cold War innovation, and British historian C.N. Hill has made similar arguments about the United Kingdom's Cold War military enterprise.[4] Others have examined the Soviet Union, Germany, Japan, and Australia, all of which demonstrated similar relationships between military enterprise and technological change during the early years of the Cold War.

Canada's experience during this period was no different from that of its political and military allies. The emphasis on conceptual design and development of weapons and technology, which dated from the Second World War, demonstrated a fundamental shift in the government attitude toward defence science, a field that had been largely ignored before the war. Although the National Research Council (NRC) had undertaken limited activities before 1939, Canada's scientific and technological community was largely dispersed, underfunded, and isolated from the concerns of national defence.[5] The Department of National Defence (DND) had neither the personnel nor the finances to investigate even simple advanced technology or weapon systems, and Canada's defence policy and doctrine of the period created little requirement for large investments into research and development of scientific projects. Suffice it to say that Canada prepared for the Second World War with largely the same naïveté that had characterized its preparation for the First World War.

Still, the totality of the Second World War inexorably and fundamentally revised this attitude. Canada's National Resources Mobilization Act of 1940 had, among other things, called for the country's scientists and engineers to commit their professional services to the state in the struggle against fascism. The NRC was given the task of organizing defence science, and technology experts were recruited across the country to develop the tools of modern warfare. By and large, after struggle and evolution, the NRC succeeded in gaining an international reputation in the fields of radar, explosives, war medicine, rocketry, ballistics, and chemical and biological warfare.[6]

The war also permanently transformed the relationship between Canadian government, science and technology, and defence. The total mobilization of science and engineering between 1939 and 1945 resulted in the elevation of several technocrats into influential decision-making positions and established a precedent for collaboration and direction that lasted well beyond the end of hostilities. The Cold War and the threat of a third hot war merely expanded the range of national security concerns and the need for science to confront them. The claim that military preparedness for a nuclear Armageddon depended on massive resources and increased federal support for defence science proved far more convincing than ever before. Canadian scientists, once ostracized from defence budgeting and decision making, suddenly found themselves at its very centre. It was from this vantage point that a Canadian rocket and space program was born.

Though Canadian interest in researching the upper atmosphere dated from the nineteenth century, a government-driven rocket and space program began only after the Second World War. Dozens of university space research ventures were initiated in the late 1940s, and the newly created Defence Research Board (DRB) sponsored the design and construction of the first long-range missiles and, later, sophisticated upper atmospheric launch vehicles for both military and civilian applications. In 1955, in cooperation with the US Department of Defense (DOD), the Canadian Army established a permanent launch facility at Fort Churchill, Manitoba, near the southwest coast of Hudson Bay, where the Black Brant rocket as well as several American-built rockets and missiles were tested and employed. At the same time, the DRB initiated discussions with the DOD for Canadian participation in US ballistic missile defence research and development. By October 1957, when the Soviet Union shocked the world with the launch of Sputnik, the first man-made object in space, Canada's plans for its own first satellite project were already well advanced.

Canada's space program reached a zenith during the 1960s. In late 1962, albeit with US aid, Canada launched its own satellite, named Alouette, into orbit.[7] It had also developed critical components for several American space assets and was a key contributor to US manned spaceflight systems. Canada's vital partnership with the United States in North American Air Defence (NORAD) was testimony to its ability in contributing to the deterrence of a Soviet attack. Furthermore, the deployment of Canadian surveillance of space technology provided some of the earliest assessments of Soviet intercontinental ballistic missiles (ICBM) and space-based systems. In addition, DND was engaged in a number of ballistic missile defence projects and counter-space-operations studies, and had solidified working relationships with a number of defence space and intelligence agencies in the United States. This was topped with a super gun launcher program at McGill University, a second Alouette satellite launch in 1965, and the completion of Canada's first rocket development program in 1967. When Canada's space effort finally paused near the end of the decade, a legacy of achievement during the golden age of space exploration, largely based on security and defence, was clearly identifiable.

Yet, as with much of the history of Canadian science and technology, the details of the origin of the Canadian rocketry and space program remain elusive. Historian Douglas Owram devoted only a single page to the history of science and technology in his omnibus examination of Canadian historical literature, most of the entries identified being associated with the history of medicine. Likewise, Carl Berger's study of the writing of Canadian history devotes a single page to the same subject, noting that "literature on the history of science in Canada has concentrated on those features of it that were comprehensible to historians who lacked any formal training in the field."[8]

This may seem somewhat surprising given the considerable literature that exists on US-Canada defence cooperation during the Cold War. However, most if not all Canadian studies on this subject portray the relationship as that between only two actors – defence and government. This study argues that a third actor, the scientific and technological community, also figured in the association and more importantly played a critical part in shaping the evolution of both that relationship and Canada's role in strategic defence with other Western allies.

Such an obvious and biased gap in Canada's space history merits further attention, not only to answer questions about Canadian post-war defence science and space technology but also to identify important aspects of the Cold

War Canada-US relationship not addressed in the literature. The formative years of the Canadian space program paralleled, albeit on a much more modest scale, the evolution of the US space program. The bilateral defence relationship between the two countries was strengthened by the need for cooperative defence in the Cold War security environment and led to the creation of such missile- and space-oriented organizations as NORAD.

Similarly, Canada influenced several American space projects, both civilian and military, and a number of Canadian scientists and engineers took leading roles in American agencies such as NASA.[9] Participation in bilateral space activities shaped Canada's post-war national security, science, technology, and industrial policies. Canada also shared similar experiences in its quest to modernize its post-war defence and to then diffuse that technology into its civilian economy, albeit with less success. The outcome was a comparative yet different linking of national defence to national welfare during this period. These issues alone make the study of Canada's rocketry and space program critical to investigating Canada's Cold War interests and its bilateral security relationship with the United States.

Defence and Discovery is organized along both thematic and chronological lines and focuses on the period of Canada's security-driven space science research and technological development between 1945 and 1974. The book situates the reader within the history of the Canadian space program and its various projects, providing a framework of knowledge for further scholarship and debate. It also analyzes the connection between Canadian science and technology, defence, and government, the actions taken, and their impact on the execution of the country's space program during the golden age of exploration and exploitation.

Overall, this book seeks to reveal the nature of the first three decades of Canada's exploitation of space and to place it within the wider context of the political history of Canadian science and technology. Although the United States, Russia, Britain, and other nations are aggressively recording their space exploration legacies, Canada has yet to embrace, officially or otherwise, telling the great tale of defence and discovery that launched it into outer space.[10] This work seeks to end that status quo by demonstrating that Canada was an active and at times critical element of the Cold War space race.

1
Cold War Security and Aerospace Defence Research Prior to Sputnik

In Canada the study of aerospace-related disciplines such as astronomy and physics may be traced as far back as the eighteenth century, but the direct study of the upper atmosphere and outer space began much more recently.[1] In 1882 and 1883, Canadian scientists participated in the first internationally coordinated study of meteorological, magnetic, and aurora phenomena in northern Canada, as part of a program known as the International Polar Year (IPY). A second IPY was held in 1932, and another generation of Canadian scientists and explorers using new radio techniques successfully conducted Canada's first measurement of the ionosphere.[2] These techniques were improved during the Second World War, as military imperatives provided massive funding and resources to previously strapped research and development in ballistics and space sciences. By the end of the war, sustained government support for national-level science programs had paved the way for an ambitious foray into upper atmosphere/space technology research and development under the aegis of defence-related priorities.[3] During the Cold War, US support and the threat posed by an unstable and potentially hostile Soviet adversary assured Canada an early entry into the exploration and exploitation of outer space.

Science, Defence, and Government

As the Cold War descended into its darkest period, Canadian defence scientists and engineers became acutely aware of their increasingly important role in deciding its outcome. Speaking before an audience in early 1955, J.J. Green, a senior analyst with the Defence Research Board (DRB) of Canada, noted that "no one can foresee how history will judge this century but it is not too difficult to put down on paper some of the things for which we shall be remembered. Among the more important of these, the men and women of the future will, I am sure, record that our generation were the first to apply science to warfare on an organized basis."[4] Green was more correct than he could have imagined, but even then, considerable effort had been required to bring Canada's scientific and technological capabilities to bear, first, in the liberation of Europe and after the war in the ongoing defence of North America.

At the end of the First World War, Canada's national science program remained largely undeveloped. The annual expenditure on government laboratory research was approximately $1 million, and of the nearly twenty-four hundred leading Canadian firms, less than forty had laboratories. In industry, less than $150,000 was spent annually on research and development.[5] Though the National Research Council (NRC) was officially created in 1916 to direct national scientific research and development, its early post-war years were plagued with rivalry and competition from Canadian universities for scarce government funding.[6] The situation improved somewhat during the interwar years, with a notable but limited increase in federal funding and of scholarships tendered through Canadian universities by the NRC.[7] Scientific manpower also increased slowly during this period, though a number of Canadian scientists continued to travel to the United States in search of gainful employment. In general, though interested in scientific development, the federal government remained reluctant to devote significant resources to research beyond those already specified in the NRC Act of 1924.

The detached relationship between science and government in Canada changed drastically during the Second World War. The National Resources Mobilization Act of June 1940, designed to concentrate all of Canada's resources on the defeat of the Axis powers, essentially congregated nearly all of Canada's scientists within the NRC and its ancillary departments. There, they were provided with funding and resources unlike any previously received from Ottawa and were able to rapidly expand the country's scientific capabilities and post-war potential for research and development. Between 1938 and 1945, government research and development expenditure increased sevenfold, from $4.9 million to $34.5 million, or roughly 0.3 percent of gross national expenditure.[8] This outlay decreased somewhat at the end of hostilities, yet, thanks to defence, the massive influx of funding and effort into centralized research and development forever altered the traditional relationship between science and government in Canada.[9]

The war also created an environment that allowed prominent Canadian scientists and engineers, such as NRC president C.J. Mackenzie, to join the upper ranks of the decision makers in Ottawa. A close friend of the legendary C.D. Howe, minister of trade and commerce (and later director of Canada's war munitions and supply), Mackenzie exerted an influence over the direction of Canada's national science and technology efforts during and after the war that members of his profession had not previously enjoyed. Other prominent researchers of the period included John Cockcroft, wartime scientific director of the Anglo-Canadian atomic projects at Montreal and Chalk River; E.G.D. Murray, director of Canada's biological warfare program between 1941 and 1945; Colonel Omond Solandt, a leading expert in Second World War operational research and medicine, and later the first director general of Canada's post-war

Defence Research Board; and George Wright, a major influence in Canada's wartime explosives and propellants programs.

Although the connection between Mackenzie and Howe was not equal to the apparently close relationship between Winston Churchill and his wartime science adviser Frederick Lindemann, there is little question that Mackenzie's regular access to Howe and the innermost circles of wartime government allowed him and his fellow scientists and engineers to enjoy an unprecedented position of power and status within Canada's scientific community.[10] One example of this may be found in the NRC annual report for 1944-45. Almost twenty years after its inception, the NRC explicitly noted for the first time that one of its major functions was to act as "advisor to the various departments of government, particularly those of National Defence, Munitions and Supply and Reconstruction."[11] Such was the result of gaining access to the inner circle of government.

Yet, as the tide of the Second World War turned in favour of Canada and its allies, it remained uncertain whether the end of the conflict would return Canadian scientists to isolation or if they would remain heavily involved in post-war nation building. Any longevity for national scientific research depended greatly on the creation of a solid post-war science policy for Canada. Realizing this necessity, the government, the military, scientists, and industrial leaders met even before the end of hostilities to consider the options. In February 1944, C.J. Mackenzie presented a paper at the Engineering Institute of Canada in which he outlined four basic requirements for a successful future national science program.[12]

First, Mackenzie argued that it was essential to retain the best Canadian scientific personnel within the country or at least the Canadian establishment. Second, the government must continue to expand its financial support for national research programs; if career paths were to remain competitive, salaries and wage structures for scientists, engineers, and technicians required increases and reorganization. Industrial research needed to be encouraged as well, potentially through liberal tax policies and reimbursements for technological developments made available to the public. Third, if the above were to succeed, greater coordination was necessary between the various branches and departments focused on Canadian research and development. Lastly, once the war ended, Canadian research would need to be refocused from a purely military application to a balanced peacetime program of both defence and civilian interests. If all this could be accomplished, Mackenzie stated, Canada might become a world leader in post-war science and technology.[13]

In mid-1944 the Department of National Defence (DND) also began meeting internally to consider the issue of post-war defence research in Canada. It was obvious that, at war's end, a large portion of the military scientific corps would request a release from service and seek civilian employment elsewhere. To ensure that some form of defence research program existed in the post-war period, Air Vice-Marshal E.W. Stedman, the director general for air research, recommended that a permanent Cabinet committee on defence research be formed under a chairman nominated by the government. The idea was endorsed by the Cabinet War Committee and later approved by Cabinet on 3 October 1944.[14] The committee, which was to be chaired by C.D. Howe, would include the three armed service chiefs of staff, the president of the NRC, two representatives from industry, and two civilian members. Cabinet again concurred and approved the terms of reference for the committee on 10 August 1945.[15] It was expected that, after an initial period to consider the issue, the committee would meet near the end of the year.

Ten days after the approval of the committee's terms of reference, Lieutenant General Charles Foulkes was appointed the chief of the general staff.[16] A career infantry officer with the Royal Canadian Regiment, Foulkes had joined the army in 1926; he went on to command the Second Canadian Division and, briefly, the Second Canadian Corps in Normandy. He later commanded the First Canadian Corps in Italy through to the end of the war. His appointment was merited but a bit surprising to some at DND who assumed that, after the war, Canada's top military command would go to the more charismatic and operationally successful general Guy Simonds. Foulkes was not a personable character, but he was efficient, an attribute that probably earned him the position.

Foulkes also clearly understood the impact of science and technology on defence. He required no convincing that peacetime defence research and development were crucial to future Canadian defence planning. Just over a week after he assumed office, he directed his staff to prepare an appreciation of the organization required for defence research in the early post-war period. His immediate concern was that, if defence research were left to the three services, duplication of effort, inter-service rivalry, and reduced post-war defence budgets would seriously diminish any capability. Simply, Foulkes felt that Canada's military scientific effort was too vital to be squandered by inter-service competition.[17]

The creation of a military division within the National Research Council, with an NRC vice-president assigned as its director, was one way of removing defence research from the control of the three services. Foulkes discussed this option with C.J. Mackenzie, but he discovered that the NRC president was reluctant to accept responsibility for Canada's post-war defence research. Although the NRC had assumed this duty in wartime, Mackenzie was anxious to return its focus to fundamental research and the civilian sector.[18] He also argued that, though the NRC had substantial resources, the financial and administrative responsibility for military research was large enough that it ultimately required its own dedicated staff.[19] Mackenzie's advice was accepted, and though neither man could have realized it at the time, the decision was a blessing in disguise. By the end of 1944, the NRC had become riddled with a small number of Soviet spies as well as other security leaks, and their unfettered access to other sections of DND might have proved disastrous for defence research and security.[20]

The end result of the Foulkes-Mackenzie meetings was that neither the armed forces nor the NRC seemed the appropriate vehicle for directing post-war defence research. Another solution was needed. Further analysis from within DND produced a number of options, the most provocative of which was a recommendation from Colonel William Wallace Goforth, head of the Directorate of Staff Duties (Weapons) at Army Headquarters, for a new defence research agency led by a representative at the chief of staff level. Goforth's proposal found immediate favour with the chief of the general staff, who passed it on to his three service chiefs. They too agreed to the concept in principle, though the chief of the air staff, Air Marshal Robert Leckie, had reservations concerning the plan. With a general consensus, however, Foulkes prepared a memorandum outlining the proposal for submission to the Cabinet Committee on Research for Defence.[21]

Surprisingly, the committee met only once, on 4 December 1945, in C.D. Howe's Parliament Hill office. Among those present were Minister of National Defence Douglas Charles Abbott, the secretary of the Privy Council Office, Wing Commander A.M. Cameron, C.J. Mackenzie, and Foulkes and a number of his military staff.[22] Foulkes's memorandum was reviewed in detail, with those present agreeing that research for the three services would be coordinated under a single director general for defence research and that this person should be a civilian with scientific training.

Further, since this course of action simply required reorganization within the department, no Cabinet approval was needed to implement it. All that was necessary was the passage of an Order-in-Council authorizing the appointment of the director general for defence research. A note was made to have the order drafted. Then, a subcommittee composed of Mackenzie and the chiefs of staff was formed to generate a list of potential appointees. It was expected that the position of director general would be filled in short order, and with that, the future of defence research in Canada would be all but determined. With nods and smiles all around the room, C.D. Howe adjourned the committee meeting.

But before everyone left his office, in a show of bureaucratic flexibility and brevity that is seldom observed today, someone suggested that a suitable candidate for director general might be Colonel Omond McKillop Solandt, a senior Canadian operational research analyst who had recently been appointed chief scientific adviser to Lord Louis Mountbatten. A graduate of the University of Toronto, Solandt was both a scientist and a qualified medical doctor who had completed his graduate studies at Toronto's Banting and Best Department of Medical Research.

In 1938 Solandt went to Cambridge, where he worked under Sir Joseph Bancroft. When the war began, Solandt held several senior appointments in medical research in England. Later work dealing with physiological problems related to tank personnel brought him to the appointment of deputy director of the British Army's Operational Research Group, which he then commanded beginning in May 1944. In early 1945, Solandt was appointed to Mountbatten's staff but was soon assigned as a member of the Joint Military Mission sent to Japan to evaluate the effects of the atomic bombs used on Hiroshima and Nagasaki. He returned to England later that year after completing his research. Not a career soldier, he was expected to resign from military service at the end of the war and return to civilian employment. However, his education, background, and experience in both administrative duties and wartime operational research made Solandt an excellent candidate for the position of director general.[23]

The armed service chiefs, who were still in Howe's office, concurred with the recommendation, and a conversation between Foulkes and C.H. Best at the University of Toronto later that afternoon confirmed the nomination. Ten days later, on 14 December 1945, the minister of national defence submitted the draft Order-in-Council for the reorganization of research and the appointment of a director general for defence research; the Privy Council officially approved his request on 28 December. Solandt was then informed of his new assignment, and arrangements were made for his immediate repatriation to Canada. The first phase in the creation of the Defence Research Board was complete.

Science and Security

By the end of the Second World War, science and technology had delivered to the Western allies, among others, the "winning weapon," as it were, in the form of the atomic bomb. Yet even before the end of hostilities, the supposed wartime ally, the Soviet Union, had stolen the recipe for this device and many other Western technological secrets. These actions rapidly degraded the US strategic advantage over the military power of its increasingly acrimonious Soviet ally.[24]

More often than not, the Soviet Union targeted scientists and engineers in its efforts to extract technological secrets. Smart, innovative, and conscientious, many of these people working in wartime defence projects feared the potentially destructive power they had developed and worried about its illegitimate use in the future. Made vulnerable by their feelings of guilt, they sometimes acquiesced to Soviet agent encouragement that they "share" their knowledge with the USSR in the false hope that a balance of power would be established in the post-war world. This proved ineffective in creating a cooperative post-1945 scientific environment, and whatever the intention, it was disastrous for Western security, degraded the West's military technological lead over the Soviets, and ultimately placed the entire Western allied scientific community under many years of close surveillance, suspicion, and prosecution by government.[25]

Canada's scientists were not excluded from this situation. The defection of cipher clerk Igor Gouzenko from the Soviet embassy in Ottawa on 5 September 1945 and his subsequent detention and interrogation revealed that a number of Canadian public servants, including some prominent scientists working with the NRC, might have been involved in spying for the USSR. After accepting Gouzenko's account as true, Canadian officials deemed the situation so serious that Norman Robertson, the under-secretary of state for external affairs, immediately reported the defection directly to Prime Minister William Lyon Mackenzie King.[26]

King displayed particular interest in Gouzenko's alleged security leaks that related to breaches in Canada's defence research. As he noted in his diary on 7 September 1945, Robertson had reported "that everything was much worse than we could have believed ... In our Research Laboratories ... where we had been working on the atomic bomb there is a scientist who is a Russian agent. In the Research Laboratories in Montreal ... there is an English scientist who is ... acting as a Russian agent."[27]

C.J. Mackenzie was immediately notified of the spies in the laboratories, and security experts were brought in to investigate suspects now under surveillance.[28] Arrests were made in February 1946, followed by the convening of a Royal Commission on Espionage and a series of treason trials against various individuals whose names, for one reason or another, appeared in the documents that Gouzenko brought with him to the Canadian authorities.[29] Several other scientists who were not named in the Gouzenko papers were put under surveillance as distrust and paranoia suddenly swept through the ranks of the Canadian scientific research community.

In addition, a number of organizations, such as the Canadian Association of Scientific Workers (CAScW), were investigated by Canadian authorities. Defence scientists associated with such groups often had their security clearances "temporarily" revoked or were simply denied any new clearance and access to restricted projects. Membership at the CAScW dropped dramatically as did that in other unions. Scientists and engineers who did not belong to these associations were encouraged not to join if they wished to retain their employment. For organized defence research labour, the Gouzenko affair proved disastrous.

The history of these investigations and their outcomes is already well explored in academic literature, but it is important to note here that the scientist spy scare and the potential theft of restricted defence technology emphasized two critical points for the Canadian government.[30] First, as historian Donald Avery states, "by the summer of 1946 Canadian military planners no longer considered war with Russia a remote possibility"; furthermore, they agreed that such a war would probably be won by whoever had the best long-range bombers, rockets, and atomic bombs.[31] As early as 10 September 1945, even Norman Robertson and Prime Minister King conceded the grave possibility that any future war between the United States and the USSR would probably come through Canada. There could be no sitting idly by if such a conflict arose.

Second, given that defence technologies were the likely keys to victory in a future conflict, Canadian national security must encompass the organization and protection of strategic research and development.[32] The government directed its law enforcement agencies to respond to the demand for better security as threats became increasingly apparent. Meanwhile, bringing the organization of defence research and development under the umbrella of DND soon became a foregone conclusion.

Creating the Defence Research Board

Although it was understood that scientific and technological innovation was crucial to the development of Canada's post-war defence and security, transforming this idea into reality required a serious commitment. Interestingly, this commitment did exist in two men who were critical to the process: Omond Solandt, the man chosen to become director general for defence research, and Brian Brooke Claxton, the incoming minister of national defence.

As Canada's first post-war defence minister, Claxton was under tremendous pressure to effect the demobilization of the country's massive wartime armed forces while at the same time ensuring that future security requirements were addressed. Fortunately, Claxton, a First World War veteran, was no stranger to the environment covered by his portfolio. After he served with the Canadian Field Artillery on the Western Front, his political career took him to oversight of national health and welfare issues. Claxton felt that people mattered the most and that their innovation would transform defence research and development.

Both Claxton and Solandt understood that scientists and engineers would have little influence in Canadian Cold War military modernization unless

DND created internal positions in which they could serve as advocates for new concepts. Further, the effective management of defence science and technology required leadership and authority similar to that found in the admirals and generals who commanded the armed services. Despite orders from the prime minister to get tough with the department and begin the massive postwar reduction of Canada's armed forces, Claxton remained committed to ensuring that defence research received proper support and funding, was provided with appropriate leadership, and was adequately organized and staffed. Senior-level support was critical to getting past the many obstacles placed in the way of creating a DRB.

Canada's assured access to outer space so early in the Cold War would have been impossible without its investment in defence science and technology. The tools needed for space flight and research were essentially converted weapons of war. Missiles carrying warheads became rockets carrying payloads. The defence research community also had the political and financial means to exploit this technology and transform space research from concept to reality. The DRB was therefore the only organization in Canada that could create the conditions for success; it could advance Canada's own goal to explore and defend the upper atmosphere and outer space as well as make a meaningful contribution to Cold War missile, rocketry, and space programs in the United States and elsewhere.

While the NRC struggled to extricate itself from the growing scientist espionage crisis, Omond Solandt began laying the foundation for the organization and activation of the DRB. From modest temporary accommodations on Slater Street in Ottawa, Solandt and his tiny staff of four, including a First World War veteran from the Veterans' Guard of Canada, compiled initial estimates for budget and staffing requirements, as well as a plan for the review of all existing and contemplated research projects from which recommendations would be made concerning the direction of the research effort. In February 1946, Solandt submitted a request to the Treasury Board for an initial budget of $1 million for defence research within the War Appropriations of 1946-47, over and above the request for $14 million already submitted by the research and development programs of the three services. The funds were to be largely invested in creating the initial facilities and staff with which to complete a survey of Canadian defence research during that year.[33]

Like any new organization, the DRB suffered endless growing pains as DND began reallocating ownership of defence research to it. From the outset, the DRB was faced with complications. The plan to spend all of 1946 in completing a survey and assessing the status and future direction of defence research in Canada was abruptly cut short when the Cabinet Committee on Defence made a hurried request to Solandt for a comprehensive policy paper on the subject no later than the end of April. The sudden announcement of an informal conference on Commonwealth defence science in London, England, during June 1946 demanded that Canada be prepared to table some form of policy during this event, but it had not yet even agreed on the basic principles such a policy might contain. Regardless, there was little time to spare; on 17 April 1946, after a concentrated effort, Solandt produced a draft paper for review titled "Policy and Plans for Defence Research in Canada." The document was discussed in detail at a Chiefs of Staff Committee meeting on 30 April.[34]

The first draft met with a generally cool reception from the committee, particularly from the chief of the air staff, Robert Leckie. A First World War veteran, Leckie had served as director of civil flying operations in Canada and as the commander of the Commonwealth Air Training Program during the Second World War. Appointed chief of the air staff in 1944, he was intensely loyal to the Royal Canadian Air Force (RCAF) and protective of its research and development programs.

Leckie and his technical staff were concerned that the proposed DRB organization was too ambitious and that its authority over each service's research and development programs would be too great. Although it was already agreed in principle that the DRB was to become responsible for all defence research and development, the three services attempted to stake new claims over their respective territories. As DRB official historian Captain D.J. Goodspeed later noted, the RCAF had a particular agenda, already planning for the development of jet engines and fighter aircraft designed and built in Canada.[35] A DRB with comprehensive powers and budgetary authority would interfere with this ambitious post-war agenda.[36]

At the conclusion of the meeting, it was agreed that a revised draft acceptable to all parties should be prepared as soon as possible. Solandt returned with a new paper the next week, yet only a portion of it was unanimously approved.[37] The document was amended, and a final version was approved and forwarded to the Cabinet Committee on Defence shortly after 14 May.[38]

Divided into three parts, the DRB Cabinet policy paper emphasized a number of key points that shaped the future of defence research in Canada. First, Solandt was firmly opposed to forming a static organization that would rotate around Ottawa and its bureaucracy. Although he acknowledged that Canadian efforts were not large enough to warrant separate establishments for research, development, and production, he recommended that DRB facilities in which these were combined should be located throughout Canada to best take advantage of its geographical situation. The choice proved both very popular and successful. As Solandt later reported, "When we were planning the DRB establishments after the war we definitely decided we were going to put them right across Canada, and looking back now it was a very wise decision."[39]

The policy document stressed other points as well. The DRB estimated that, because general war with the Soviet Union or its allies was at least a decade away, it could focus on long-term research and development without concerning itself too greatly with immediate requirements. And though the intent was to develop a completely independent defence research capability in Canada, the initial plan was to concentrate, not on everything, but only on those areas in which Canada had important original ideas or special interests, facilities, or resources to devote to a research problem.[40] To retain access to information on research in which the DRB was not engaged, Canadian military and civilian personnel were attached to other defence research facilities in both the United States and the United Kingdom to act as liaison officers and observers. The chief of the general staff – who felt that, were a major war to arise, Canada would fight alongside either its American or British allies – readily accepted this part of the policy.

Though accepted and ratified, a final part of the policy paper encountered strong opposition from the air staff.[41] In it, Solandt recommended that the DRB concentrate primarily on research, limiting its involvement in the engineering, design, and development stages of defence projects. Further, he suggested that the actual design and construction could take place outside of Canada. This point in particular upset the RCAF senior leadership. During the Second World War, Canada retained limited control over its aircraft design and development and had been unable even to secure Canadian-manufactured Hurricane aircraft for its own defence. As a result, having control over its own aircraft research, development, and procurement had become an article of faith for the RCAF.[42] Leckie and his successors, who had every intention of controlling the research and development for their next generation of aircraft, were reluctant to turn it over to a new Canadian defence research organization, let alone a foreign agency. Any thought of foreign control was anathema to their agenda.[43]

This highly contentious issue and its overall implications for the long-term health of the DRB deserve further consideration. The DRB official history and subsequent recorded interviews with Solandt by scholars suggest that his motive for pushing this policy was to prevent defence researchers from becoming immersed in the administration and bureaucracy of seeing a project through to completion. By allowing his scientists to focus on the fundamental aspects of research, choosing a small number of long-term projects that might take years to reach the design stage, he could provide them with an environment in which to conduct pure research rather than application. Although it was certainly favourable to the scientists themselves because it enabled them to pursue their own interests, this decision was an ill-planted seed that, twenty-five years later, would blossom into accusations of "irrelevance" and "goal displacement" from the federal Royal Commission on Government Organization. Ultimately, the DRB's intent to restrict its efforts to the purely conceptual and research stage contributed to its untimely disbandment.

The appendix of the policy paper suggested nineteen specific fields of research, including guided missiles, rockets, and meteorology. The first two of these fell within the domain of the army's armament research and development establishment, and the latter included investigations into the nature and composition of the upper atmosphere such as those projects already in place under Royal Canadian Navy (RCN) oversight at Fort Churchill. Like its allies, Canada sought knowledge on both rocketry and upper atmospheric conditions early in the post-war period, realizing the importance of both in the development of advanced weapons and defence systems. Canada became particularly proficient in the latter, so much so that the United States curbed its own efforts in the field and relied largely on Canada for data and information.[44]

The Commonwealth Conference on Defence Science, held in London, England, in June 1946, marked the first public announcement of Canada's post-war defence research policy. Despite being overtaken by events, Solandt and the Cabinet Committee on Defence had created a sound baseline document in a very short time that was well received by Canada's allies. At home, the new policy laid the groundwork for the creation of a permanent DRB staff and for the transfer of army, navy, and air force research and development establishments to its authority. Near the end of the year, the DRB had established its Personnel Selection Committee under the leadership of Otto Maas, Canada's wartime director of chemical warfare and explosives programs, and had made arrangements to retain the services of other valuable scientists already working at the Department of National Defence. Previously employed by McGill University's Chemistry Department, Otto Maas himself was appointed director of the DRB biological and chemical warfare research division in 1947.

Finally, the chief of the general staff initiated the legislation required to revise the 1927 National Defence Act to include the Defence Research Board. Despite continued resistance from the RCAF, the issue was put before the newly appointed minister of national defence, Brian Brooke Claxton, on 20 January 1947. A final bill designed to amend the 1927 act so as to include the formation of the DRB was introduced into the House of Commons on 7 February. The bill passed through the required three readings without great discussion or debate, and on 28 March 1947 the amendment became law. The Defence Research Board had achieved legal status and authority.[45]

The initial organization of the DRB clearly reflected its mandate and intent to become the centre of gravity for Canadian defence science and technology research. Under the chairmanship of Omond Solandt, the board was initially a loosely organized collection of advisers, laboratories, sections, stations, and research establishments scattered across the country and tied to any one of the three armed services. As an equal to the chiefs of staff and adjunct to the deputy minister of national defence, Solandt was supported by two deputies – a deputy director general and an administrative deputy. His immediate counsel also included a secretary of the board, scientific advisers from each of the three services, a project coordination section, and liaison offices in London and Washington. The DRB chairman also oversaw the Directorate of Research Personnel and the Directorate of General Services, both of which supported the many other groups under his stewardship.

A number of organizations formed the body of Canada's post-war defence research network. Existing facilities included the army's Canadian Armament Research and Development Establishment (CARDE) at Valcartier, Quebec, the Suffield Experimental Station in Alberta, the Defence Research Chemical Laboratories in Ottawa, the Kingston (biological warfare) Laboratory, the Radio Propagation Laboratory in Ottawa, the Defence Research Establishment at Fort Churchill (renamed the Defence Research Northern Laboratory in 1947), and the Naval Research Establishment at Halifax. Other cells included the Weapons Research Section, the Electrical Research Section, the Special Problems Research Section, the Biological Research Section, the Naval Research Section, and the Scientific Intelligence Section. Additionally, the DRB formed or co-chaired a number of advisory defence research and development committees between Canada's own armed services, as well as cooperatively with allies.

When the DRB was formed, its various organizations employed roughly a thousand personnel, including two hundred scientists and between thirty and forty technical officers. In 1947 and 1948, approximately $4 million was expended on research and development, an amount that quadrupled in the years 1948-49, when materials, supplies, equipment, and salaries were added to the budget. Defence research expenditures continued to increase between $7 and $10 million a year through to 1956 and 1957, a definite indication of Canada's strong commitment to defence science during the early part of the Cold War. More importantly perhaps, the creation of the DRB clearly signalled Canada's intent to take defence research seriously and make a solid contribution to its own strategic interests as well as those of its allies.[46]

Cold War and Hot Physics

Canada worked aggressively to safeguard its post-war sovereignty with a robust and self-focused national security policy. Although it certainly supported the ideals of multilateralism and collective security through the newly created

United Nations, it also ensured that its own interests were being met through the establishment of regional or bilateral security arrangements with its allies.[47] Canada joined other Western nations to form the North Atlantic Treaty Organization (NATO) in 1949 and sought to strengthen its bilateral defence and production ties with the United States throughout the early 1950s. Even increasing anti-American political rhetoric did not deter the government from solidifying a NORAD agreement with the United States in 1957 or the ultimate acceptance of American nuclear weapons into the country soon afterward. It is important to note also that the success of many of these and other Cold War security arrangements was dependent on the development and sharing of advanced defence science and technology.

Like their counterparts in countries allied with Canada, Canadian strategic planners considered how science and technology might play a role in security. Early DRB agendas emphasize those areas in which Canada's military research was strong and in which it could achieve and sustain success.[48] Although Cabinet, DND, and the DRB all understood that the implementation of a large-scale military-industrial production complex similar to that taking shape in the United States was plainly out of the question, ways were sought to maximize those capabilities currently within reach. Interestingly, the science and technology expertise included some if not all of the most dangerous elements of future Armageddon – namely, physics, chemistry, electronics, rocketry, missiles, biological agents, chemicals, and even atomic weapons.

Among the many disciplines that led Canadian defence research toward the threshold of outer space, physics played the dominant role. Here, Canada's expertise was considerable, and the country's scientists had made major contributions to allied wartime programs in ballistics, radar, and atomic energy. They were also considered world experts in the study of the effects of the upper atmosphere on radio and communication, a field that Canadian researchers had investigated throughout the twentieth century.

On 12 December 1901, Guglielmo Marconi received a Morse signal in Newfoundland from a transmitter in England, 2,900 kilometres away. At the time, no one could explain how this was possible, as no one understood how radio waves could travel over the horizon. The following year, two scientists, an American named Arthur Kennelly and an Englishman named Oliver Heaviside, suggested that the radio waves travelled great distances because of the existence of a conducting layer somewhere in the upper atmosphere that reflected them. Their ideas prompted a long series of studies and debates over the definition, origin, and nature of what for many years was known as the Kennelly-Heaviside layer and was later named the ionosphere.

The ionosphere is simply the region of the earth's atmosphere where there is a significant concentration of ions. The region has no definite boundaries; rather, it is a dynamic area anywhere from sixty kilometres to approximately ten earth radii out into space where neutral particles are ionized and subsequently combine into neutral particles again. The degree of ionization depends on a number of factors, such as solar radiation, atmospheric composition, pressure, and temperature. The ion concentration maximum, also known as the electron number density maximum, is typically located at approximately three hundred kilometres from the surface of the earth.[49]

At first, very little detailed study of the ionosphere was undertaken in Canada, but after the introduction of wireless communications, a greater comprehension of its effects became essential to understanding and predicting its impact on radio waves and wireless radio traffic. In the mid-1920s, American scientists G. Breit and M.A. Tuve produced the pulse ionosonde, a tool that revolutionized ionospheric research in the United States and Canada. Essentially a radio transmitter and receiver, it allowed the user to deduce the height of the radio-wave-reflecting level by illuminating the ionosphere with short pulses and noting the time difference between transmission and the return of an echo. Also, the particular frequency was determined by the degree of ionization at the reflecting level, which, when combined with the pulse delay, provided the user with the height level in the ionosphere of that particular electron number density, as well as the frequency required to propagate through it relatively unaffected.[50]

Advanced Canadian research and analysis of the ionosphere began with the influx of federal funding and support during the Second World War. In 1941 the National Research Council constructed an ionosonde at Chelsea, Quebec, operated under the direction of Frank T. Davies, an experienced scientist then employed with the Operational Intelligence Center of the navy. Davies' tasks were to record constant measurements of the ionosphere and predict optimum operating frequencies for military communications. Canadian ships protecting convoys and chasing German submarines in the North Atlantic utterly depended on timely communications, making it essential that the link remain uninterrupted as much as possible. Davies' initial research effort was so successful that further installations were built, including one by the American Carnegie Institution at Clyde River, Baffin Island. In 1943, along with Lieutenant J.H. Meek, who was the superintendent of the NRC Radio Propagation Laboratory, Davies installed another ionospheric station at Cape Merry. By then, Davies and Meek were responsible for overseeing the training and inspection of all stations, including those in Canada staffed by members of the US military. In 1945 seven stations were located across the country.[51]

FIGURE 1 Frank T. Davies (1904-81) served in naval intelligence during the Second World War, pioneered the Defence Research Board's Radio Physics Laboratory, and was chief superintendent of the Defence Research Telecommunications Establishment from 1951 to 1969. Communications Research Centre, 69-17720.

Ionospheric research in Canada continued after the end of the war. Canada had attained considerable recognition and reputation among its allies as a leader in this field, and both the United States and the United Kingdom took a serious interest in supporting further efforts after the war. In 1946 the Radio Propagation Laboratory assumed the lead in advancing Canada's research activities, and the Department of Transport took over the operations and staffing of the existing ionospheric stations.[52] Meanwhile, three additional stations were built, one each at Baker Lake, Resolute Bay, and Fort Chimo. With the transfer of the

Radio Propagation Laboratory and its cousin organization, the Radio Wave Propagation Committee, to the Defence Research Board in 1947, Solandt recommended that the time had come for a considerable expansion of the facilities and funding then devoted specifically to Canadian ionospheric research. The decision was timely. In late 1947, the US Department of State requested a technical conference with Canada to discuss options for the expansion of ionospheric research, resulting in further recommendations from the DRB to the Cabinet Defence Committee for new and improved research stations all across the Canadian Arctic.[53]

Encouraging such an investment was grounded in very clear reasons, which are critical to understanding the desire for the advancement of upper atmospheric studies in Canada within the larger context of the Cold War. Both the United States and Canada predicted that any Soviet attack on North America would travel via the shortest route, over the North Pole. Early warning against such an attack would thus depend greatly on systems capable of detecting and identifying bombers and intercontinental ballistic missiles (ICBMs) well before they reached Canada's Arctic borders. ICBMs presented a particular detection challenge because the speed and trajectory of their flight sent them into space before they came down onto their target. The design of systems to detect any incoming threat early enough to react to it required a substantial knowledge of the atmospheric conditions that might affect the calibration and operation of the detection system. Essentially, if North America were to be defended against Soviet attack, more knowledge of the physical environment was needed.

Canada had other strategic motives for expanding its expertise in atmospheric research. The physical attributes of the atmosphere in the Canadian North, which were affected by the proximity of the North Pole, the situation of the North Magnetic Pole entirely within northern Canada, and occurrences such as the aurora borealis, presented unique challenges for technologies dependent on radio waves, particularly electronics and communications. The large expanse of Canadian territory and its relatively small population, approximately 17 million in the 1950s, implied that robust long-range communications would become essential to any defence of the country in the case of war and any development of the country during peacetime. Moreover, it was unknown what effects the upper atmosphere might have on advanced aerodynamics, on jet or nuclear propulsion, or on unmanned objects such as guided missiles and rockets. The sound barrier had only just been breached in October 1947, and the effects of supersonic flight on machines, let alone human beings, were still being discovered. Obtaining answers to these and other questions about the Canadian environment ultimately provided considerable value to the development of defence and sovereignty.[54]

FIGURE 2 The original Defence Research Board Radio Physics Lab in Ottawa, about late 1950s, where many of Canada's military space projects began. It was here that defence scientists and engineers pursued studies of the ionosphere and concepts for Canada's own satellites. Communications Research Centre, 51-RPL-0169.

Upper atmospheric studies in Canada were formalized in February 1951, with the creation of a new suborganization within the DRB known as the Defence Research Telecommunications Establishment (DRTE), which came under the superintendence of Frank T. Davies. Essentially an amalgamation of the Radio Propagation Laboratory and the Defence Research Electronics Laboratory, the DRTE joined the ranks of the ever-expanding DRB as its primary telecommunications research facility.[55] In the DRTE, atmospheric study and analysis was a central focus, as the distribution of its scientific personnel demonstrated.

The DRTE consisted of two primary suborganizations – the Radio Physics Laboratory (RPL) and the Electronics Laboratory. The former, previously the Radio Propagation Laboratory, came under the direction of James C.W. Scott and consisted of six subsections, including theoretical studies, atmospheric physics, radio prediction, and three specialized radio propagation units (high frequency, low frequency, and microwaves). The Electronics Laboratory, under the direction of J.W. Cox, consisted of five sections, including transistor research,

FIGURE 3 Canada's space pioneer. A leading figure in the development of the country's space program, John H. Chapman of the Defence Research Telecommunications Establishment points to a detail on the Alouette I backup satellite. Communications Research Centre, 62-6660.

radio warfare (known today as electronic warfare and countermeasures), components research, navigation research, and radar research.

In 1954 the DRTE underwent a modest reorganization of its radio and electronic research sections to include a third suborganization, the Communications Wing, under the direction of John Herbert Chapman, a recently hired postgraduate.[56] The son of a military officer, Chapman was born in London, Ontario, on 8 August 1921. He became interested in radio and physics at an early age, and during the Second World War, he served as a radar technician with the RCAF in both Africa and Europe. Returning to Canada to complete his studies, Chapman took a master's degree in physics in 1949 and a PhD in 1951, both from the Eaton Electronics Research Laboratory at McGill University. At the DRB, Chapman's new group augmented existing research in radar and applied propagation, as well as focusing on new communications-specific research and development.[57]

The RPL concentrated on a number of projects, giving serious attention to the analysis of auroral disturbances, such as the aurora borealis, that were particular to Canada's northern regions. Under the direction of two scientists, Raymond Montalbetti and W. Petrie, the RPL designed a set of tables to predict the potential impact of ionospheric activity on radio wave propagation for use by the navy. In addition, it tested new communications systems, studied the effects of lightning on radio waves and of birds on radar screens, and directed the research, development, and construction of Canada's first Cold War early warning defence system. Named the McGill Fence early warning line but more often referred to as the Mid-Canada Line, this system employed technology developed at McGill University in Montreal. Its construction began in 1954 and was completed in 1957 at a cost of $250 million. The line consisted of a chain of ninety-eight radar stations, mostly unmanned, along Canada's fifty-fifth parallel. It operated until its closure in 1965, when its function was replaced by newer systems.

With the RPL focused on the ionosphere, the Electronics Laboratory remained largely engaged in standardizing the communications requirements of the three services until 1951, when the initiation of the Velvet Glove air-to-air guided missile program required it to shift most of its support to this new weapons program. Although the lab continued a limited involvement in other missile- and space-related activities, the complexity of the Velvet Glove technologies preoccupied much of its effort until the project concluded in 1954.

The International Geophysical Year

Upper atmospheric research and space science advanced rapidly during the 1950s, eventually leading the Soviet Union and the United States to launch man-made satellites into space in 1957 and 1958. Although the political implications of taking upper atmospheric research to the next level were most germane for the two superpowers, Canada was a key player in the military and scientific aspects of these events. Closely associated with American efforts to gain significant knowledge of the earth's polar regions and to push space science into orbit, Canada's own efforts and cooperation were critical to US security and success at its earliest stages of space flight.

Interestingly enough, the idea of an International Geophysical Year (IGY) was born during a 5 April 1950 dinner party held in Washington, DC, to honour the visit by Sydney Chapman, the renowned British geophysicist.[58] During the evening, Chapman and his colleague Lloyd V. Berkner, a noted American radio scientist and security adviser to the US Department of State, proposed a third International Polar Year.[59] The period chosen for the event was 1957-58, as this followed the previous IPY of 1932 by exactly twenty-five years and was also near a time of maximum solar activity. However, the event was known as the International Geophysical [instead of Polar] Year so as to reflect the advances in the scientific field.

Yet there were other motivations as well. In 1950 the US government was about to release a science and foreign relations report authored by Berkner that strongly emphasized the military and diplomatic importance of the polar regions. The report made the obvious statement that "certain definite benefits which are highly essential to the security and welfare of the United States, both generally and with respect to the progress of science, stem from international cooperation and exchange with respect to scientific matters," and its classified supplement stressed that increased American contact with foreign scientists would be of importance to US intelligence collection.[60]

The primary American agency overseeing the evolution of the IGY agenda was the V-2 Upper Atmosphere Research Panel, a group of military and university interests responsible for the first US sounding rocket program. From 1950 onward, the panel held a number of meetings with American and international scientists to discuss potential IGY activities. Canadians were often invited to these early events, though their participation on committees was formalized only after an agreement was put in place to establish a large research rocket range at Fort Churchill. In 1953 the panel formed a Special Committee for the IGY to work with the American National Research Council to oversee the development of the Fort Churchill site. With assistance from the council's Technical Panel on Rocketry, plans were executed to have the US Army set up an Aerobee rocket tower and a Nike-Cajun launcher at the Defence Research Northern Laboratory (DRNL) station.[61]

In Canada, Donald C. Rose, a scientist with the NRC Physics Division and chairman of the Canadian Organizing Committee, directed IGY activities, with several projects dispersed among many research bodies and universities.[62] Both the DRB and the NRC were engaged in a number of assignments; the DRB was concentrated at Fort Churchill, and the NRC supported various projects across the country. Interesting among the latter was the work of the NRC's Upper Atmosphere Research Section, which was headed by Peter Millman, an accomplished meteor astronomer previously employed with the RCAF. Using visual observation and radar, Millman's research teams collected a range of data on meteors entering the atmosphere at high speeds. Millman discovered not only that meteor trains were persistent visually but also that they maintained "visibility" to radar for some seconds after they entered the atmosphere. He assumed that this was caused by ionization created in the upper atmosphere as the object lost material. Using these techniques in combination with radar observations of the backscatter of echoes, Millman's teams were able to ascertain

FIGURE 4 Testing an Aerobee 150 rocket. Communications Research Centre, 72-25814.

the velocities of large numbers of meteorites. Presumably, similar techniques might later have been applicable to the detection of man-made objects re-entering the atmosphere.[63]

The primary DRB objectives for the IGY included the use of a wide variety of rockets to investigate spectroscopic and ionic characteristics of the upper atmosphere.[64] Two of these rockets were instrumented by CARDE personnel to measure infrared background levels of the upper atmosphere at altitudes up to about 250 kilometres. Although these were perhaps a minor accomplishment

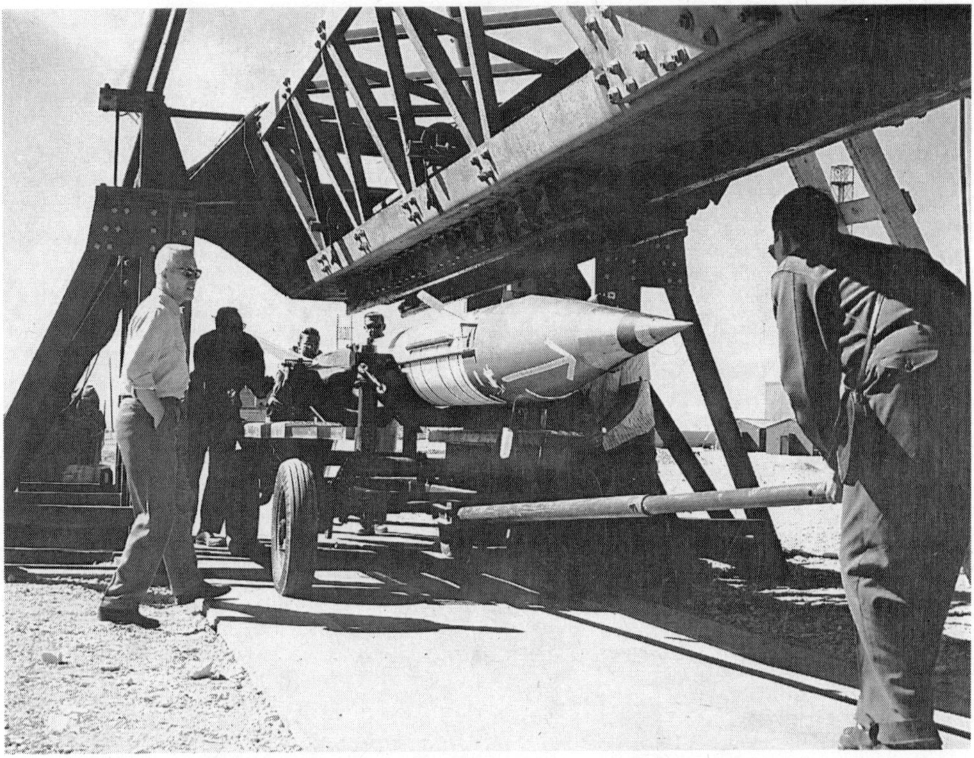

FIGURE 5 Moving an Aerobee 150 to the launch tower. Communications Research Centre, 72-25817.

compared to the American and Soviet satellites then being launched, their deployment marked the first time that Canadian instruments were carried beyond the atmosphere.[65]

The arrival of IGY activities at Fort Churchill completely altered its organization and infrastructure. In early 1956, after many years dedicated to the material testing of equipment and processes for soldiering in extreme environments, the DRNL ended its seven-year partnership with the Operational Research Group and the Biomedical Sciences Group at Fort Churchill to prepare for its new assignment of undertaking geophysical research. As these two former organizations prepared for departure, the DRNL initiated a survey of potential sites for the installation of equipment to take measurements of the ionosphere, which would be employed by Montalbetti's RPL scientists and the University of Saskatchewan. Along with the American rocket launch sites, four other sites were needed for the outlying Canadian ionosphere-measuring stations. Three of these were located south of Fort Churchill, and one was placed to the north. Two

FIGURE 6 Donald C. Rose served as the first chairman of the Associate Committee on Space Research, Canada's first organized space exploration policy group. National Research Council Archives.

were outfitted with special cameras employing the parallactic photography technique to measure the height of auroral displays, whereas the other two were equipped with an auroral recorder, a magnetometer, and an all-sky camera.[66]

Overall, the IGY was a tremendous success for everyone involved. The range of activities and the data collected contributed enormously to the nascent rocket and space programs then under way and, equally importantly, demonstrated the many possible fruits of labour that awaited innovative countries in the exploration and exploitation of outer space. For Canada, the IGY served as a validation of its emerging scientific and technological capabilities as well as an indicator of what might be achieved if it remained committed to such endeavours. Only time would tell.

The IGY had also demonstrated the high degree of cooperation built up between the United States and Canada in both space science and technology, specifically rocketry. It set the foundation for even greater Canada-US space cooperation in the years to come. Thus, as the IGY wrapped up at the end of 1958, the DRB sought to enter into formal negotiations with NASA regarding an even more ambitious ionospheric research project. The idea? To design and build the next experiments into a Canadian satellite for launch into outer space.

Standing on the Threshold of Space

Because of national security considerations and its commitment to the advancement of defence science and technology, Canada achieved considerable success in developing its upper atmospheric program during the first decade after the Second World War. From a rather conservative beginning, Canada took advantage of its post-war security situation to create the necessary means with which to research and develop space-related technologies, and to employ its limited yet highly professional resources to gravitate toward the centre of rapidly evolving Western space-related activity, usually in the United States. Further, nothing suggests that this trend was accidental: instead, an analysis of events has shown that it was a well-orchestrated manoeuvre on Canada's part, one representative of what would eventually become its modus operandi for gaining access to American rocketry and space ventures.

Canada's significance to the construction of US post-war security provided it with a considerable opportunity to advance its own rocketry and space science agenda. From the beginning of the Cold War, the United States needed access to Canadian territory and Canadian science to develop its technological response to the growing fear of Soviet Armageddon. For its part, Canada was prepared to cooperate to meet its own national security interests, but more importantly, Ottawa's commitment to post-war scientific and technological innovation in defence gave the country the means to take advantage of the evolving strategic situation. Participating in both bilateral defence cooperation and international scientific programs during the 1950s enabled Canada to demonstrate the maturity of its resources and ultimately to leverage those capabilities and its other defence contributions into negotiating its very own satellite project.

At a time when only the two superpowers had the means to break away from Earth's gravity, and only these two nations plus Britain had seriously contemplated launching satellites into space, here was Canada subtly making itself the obvious best option for the continued study of the upper atmosphere – if it had access to certain additional resources. Fortunately, these resources, such as launch facilities, could easily be provided by the US Department of Defense. As a result, Canada became an early party to the exploration and exploitation of space, something that, for most nations, remained the stuff of science fiction.

During this period, other Canadian ventures benefited from the strategy of parlaying limited resources into achieving much larger objectives. What has not been previously identified, however, is that the country employed this tactic to ensure that, at the end of the 1950s, it was well positioned to play a leading third-party role as the world stood on the threshold of space.

2

Missiles, Rocketry, and the Development of the Black Brant Launcher

In May 1949, Wing Commander R.C. Woodhead opened a secret-level meeting in Ottawa with the statement that "guided missiles were successfully used by our enemies in World War Two. If we are to be in a useful position to play a role in the event of hostilities, we too, must become acquainted and familiar with these new weapons."[1] Woodhead was addressing his colleagues on the RCAF Guided Missiles Committee, and together they were embarking on a technological path that would eventually take Canada a step closer to reaching outer space.

Contrary to popular perception, Canada officially began probing the upper atmosphere and the fringes of outer space half a decade before the launch of its now famous Alouette I satellite. Yet this less glamorous earlier period is often overlooked by historians, despite its critical role in the development of later orbital programs in both Canada and the United States.[2]

Serious Canadian attention toward upper atmospheric research came after a number of successful cooperative high-altitude ballistics and missile projects with the United States military during the late 1940s and early 1950s. Building on these achievements, the DRB accelerated its applied research in rocket propulsion to test more robust solid rocket propellants and open the door for a possible space launch capability in Canada. What developed from this effort would eventually become known as the Black Brant, arguably one of the best small payload sounding rockets ever produced.

By the time Alouette I was launched into orbit in September 1962, over thirty Black Brant rockets had already been fired, many carrying complex scientific payloads that would only later migrate onto orbiting platforms. Over a hundred Canadian Black Brants were launched by the end of 1967, in comparison to only two Canadian satellites. Thus, for the first decade of Canada's official space program, rockets, not satellites, were at centre stage of the country's efforts.

Post-War Rocketry in Canada, 1945-56

In Jules Verne's 1865 novel *From the Earth to the Moon,* a group of wayward adventurers use a massive cannon to launch themselves on their perilous journey to another world. Although the book was obviously science fiction, the science fact of the late nineteenth century assumed that being shot out of a very large

FIGURE 7 Black Brant rocket recovery. Communications Research Centre, 66-12396.

artillery piece was the only way to reach outer space. Yet, early in the twentieth century, a handful of scientists and engineers in the United States, the USSR, Britain, and Germany began to challenge this traditional concept. Three men in particular, the Russian Constantin Tsiolkovsky, the American Robert Goddard, and the German Hermann Oberth, all proposed during the 1920s and 1930s that a man would venture into space via a rocket, not through the use of artillery.[3] Their ideas, experiments, and writings sowed the seed of advanced rocketry that would soon come to fruition through war.

Prior to the Second World War, Canada displayed little official interest in the potential use of rocketry for space flight. The British Army in Canada had employed solid fuel rockets in one form or another since the War of 1812, but this did not appear to spur more focused interest over time. Instead, for Canadian researchers, the employment of rockets to reach space remained an oddity well into the twentieth century.

In fact, Canada's scientific community was generally skeptical of the many claims made by what were called rocketeers, and their ideas were challenged in

FIGURE 8 The cover of a systems manual for the Black Brant family of rockets, 1968. By the end of the test and development program, the Black Brant was ready to transition from pure defence research to civilian scientific and commercial research applications. Author's collection.

various public forums. In the April 1932 issue of *Canadian Defence Quarterly*, for example, the notable Canadian scientist and chemist Ernest LeSueur wrote, "We have in recent months been treated in the daily papers to thrilling accounts from certain, not precisely shrinking, enthusiasts as to what is to be expected from rockets and 'rocket planes,'" but he cautioned, "the average man doubtless believes that it is only a question of time before transatlantic hops will be made by rocket. Heretofore, so far as I know, none of these prophecies has been put forward by an accredited engineer."[4] In painful detail, his article continued with the many complexities surrounding velocity and gravity, summarizing rather unclearly that space flight with a rocket was theoretically impossible.

No evidence suggests that rocketry received official research support from the Canadian government prior to the Second World War. Although the NRC's Division of Physics and Engineering experimented with ballistics, investigation into rocketry or propellants was noticeably absent from the list of NRC study fields.[5] The limited historical writing on the development of science and technology in Canada during this period makes no mention of organized research in rocketry, and without further study, it is difficult to assess what occurred in this field until after 1945.[6]

In contrast, Canada's post-war DRB included a number of missile- and rocketry-related research projects among its main activities. Early efforts focused on the creation of a solid rocket fuel for Canadian-designed short-range weapons, followed later by the generation of complete weapon systems. In particular, Canada invested in the development of a new air-to-air missile for its state-of-the-art jet fighters. Named the Velvet Glove, the conventionally armed solid fuel weapon had a range of forty-five hundred metres and was originally intended for the CF-100 Canuck interceptor, which was designed and built in Canada. Approximately three hundred Velvet Gloves were completed before production was ended in 1954, when the missile was judged incapable of hitting the newest generation of Soviet bombers. Yet projects such as the Velvet Glove and others resulted in the formation of a nucleus of knowledge and experience within the DRB and allowed it to pursue larger and more complex propulsion and ballistics projects later on.[7]

From the Velvet Glove experience came an equally important project subsequently known as the Canadian Rocket Propulsion Program (CRPP).[8] Its aim was to contribute to the existing wartime defence program of applied research in propulsion, which was carried out largely by graduate science students at some Canadian universities. As well, it was expected to produce a group of experts who could assist in post-war guided weapons system studies in the armed services.[9] Finally, the CRPP intended to provide a limited production facility for rocket propellants, so that early small-scale requirements for Canadian-produced missiles could be economically met.[10] Once the infrastructure had been built, the DRB could potentially contribute to larger-scale projects – namely, the establishment of rocket capability within Canadian industry.[11]

In 1956 A.H. Zimmerman succeeded Omond Solandt as the second chairman of the DRB. He initiated a revision of the CRPP soon after taking up his new post, directing it to focus on more robust solid-state propellants that could be employed in larger rocket weapons and guided missiles similar to those used by the United States.[12] Zimmerman also moved the project from Ottawa to the

Canadian Armament Research Development Establishment (CARDE), located at Valcartier, Quebec. At the time, CARDE had several research wings engaged in all aspects of ballistics, propellant performance, missile propagation, propulsion, and active systems research (Figure 9). Perhaps most favourable to Zimmerman and the project, however, was the fact that CARDE had the experience, facilities, personnel, and sustained government funding to undertake just such a program.[13]

Zimmerman tasked the Aerophysics Wing (Figure 10) and the Explosives Wing, later renamed the Propulsion Wing under Ian R. Cameron (Figure 11), to initiate a new CRPP to improve the physical characteristics and performance of solid fuel rocket propellants.[14] Having been closely involved in several military propulsion system projects, including the Velvet Glove, these teams were an obvious choice to pioneer the revised CRPP.

Genesis: The Propulsion Test Vehicle and the Black Brant II

Though CARDE had previously developed and tested a few short-range semi-guided and guided weapon systems, it had yet to employ its newly designed propellants in any large-scale launchers. Since the aim of the revised CRPP was to achieve this goal, scientists at CARDE solved the initial problem with the design and construction of a Propulsion Test Vehicle (PTV), which was later renamed the Black Brant I.[15] The success of this rocket not only led to a new generation of indigenous launch vehicles for Canada, it put the country permanently into the exclusive club of launch-capable nations.

A number of factors influenced the conceptual design phase of the PTV. Advances in Soviet ballistic missiles and high-speed aircraft throughout the late 1940s and early 1950s informed the Canadian decision to pursue a solid rather than a liquid-based rocket propellant program. The DRB was primarily interested in developing a propulsion system that could be employed in air defence weapon roles; despite the weight constraints imposed by solid propellants, the incredibly short early warning of Soviet attack made an instant state of readiness essential and therefore this option preferable. Also, the logistical difficulties and time constraints of liquid fuel rockets, especially with regard to pre-flight fuelling, made them simply unsuitable as a rapid reaction air defence weapon.

Another important factor for Canadian rockets was stability. High performance composite rockets had to be serviceable under extreme weather conditions, such as those of the Canadian Far North. Solid propellants were generally less volatile than liquid fuels, making them much more suitable for use in the frigid temperatures of the Arctic. Because Canada's main bases were often situated in remote areas of the North, solid fuels were also considered easier to maintain.

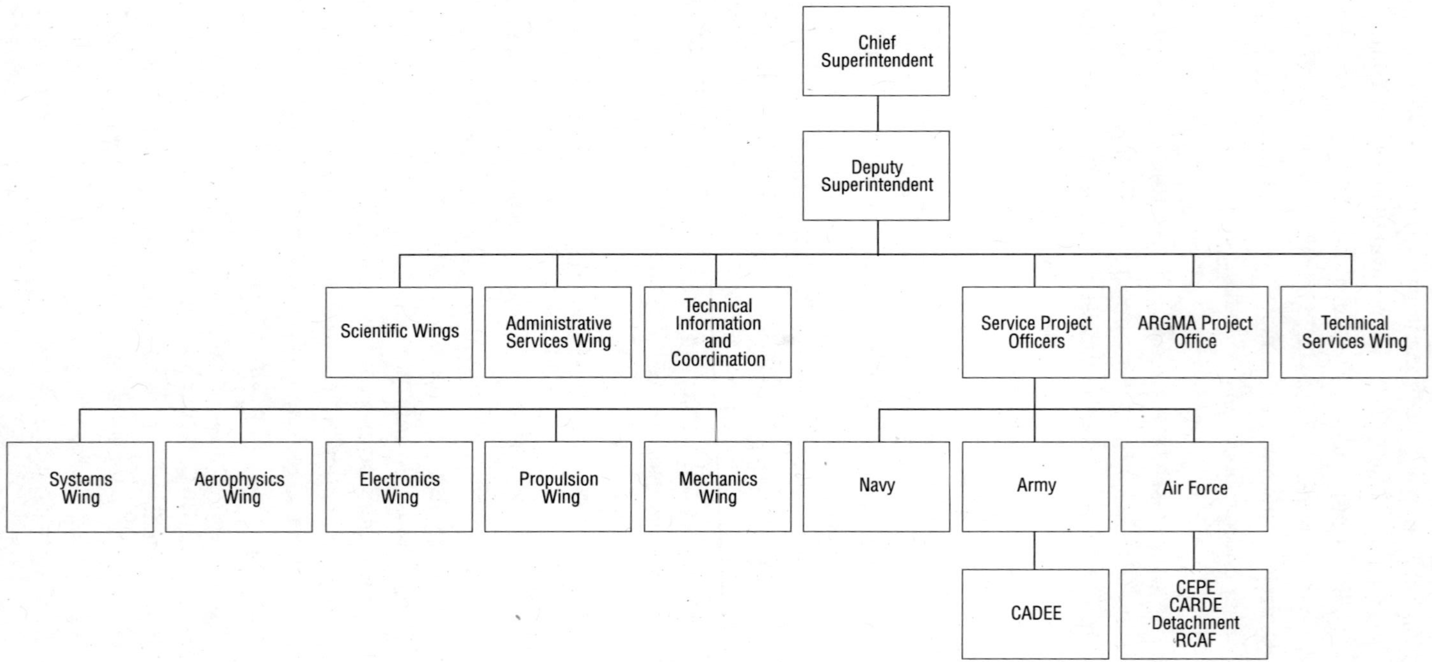

FIGURE 9 Canadian Armament Research and Development Establishment (CARDE), c. 1961. Adapted from D.J. Goodspeed, *A History of the Defence Research Board of Canada* (Ottawa: Queen's Printer, 1958), Chap. 7.

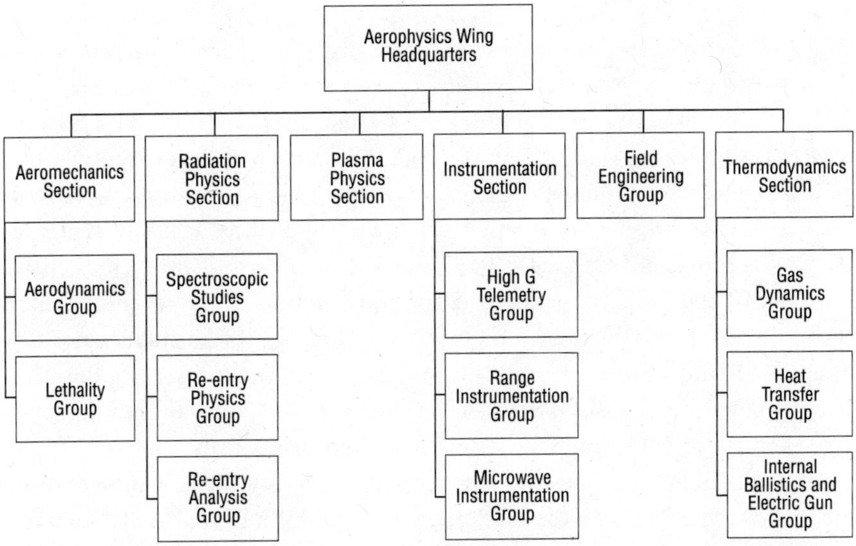

FIGURE 10 CARDE's Aerophysics Wing, c. 1961. Adapted from D.J. Goodspeed, *A History of the Defence Research Board of Canada* (Ottawa: Queen's Printer, 1958), Chap. 7.

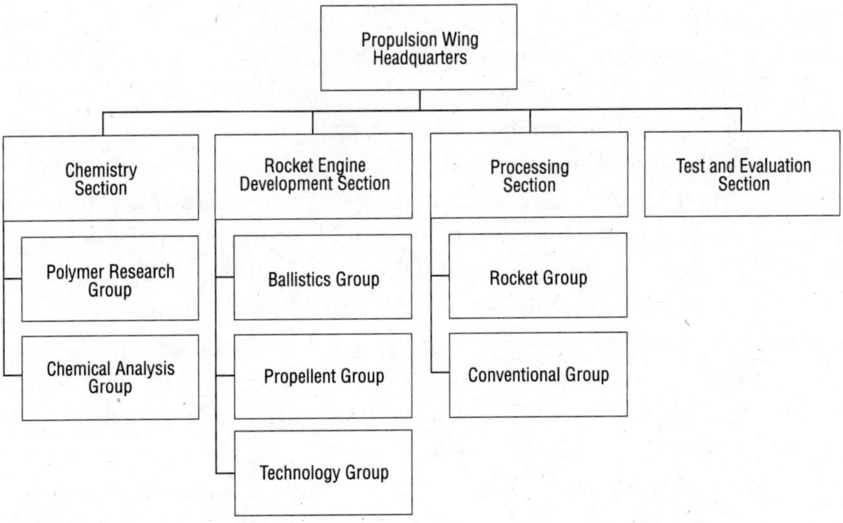

FIGURE 11 CARDE's Propulsion Wing, c. 1961. Adapted from D.J. Goodspeed, *A History of the Defence Research Board of Canada* (Ottawa: Queen's Printer, 1958), Chap. 7.

To test the solid rocket fuels then being designed, CARDE first required a PTV in which to stage the propellants. The British seventeen-inch-diameter Skylark from Bristol Aerojet of England offered an immediate prototype, so its design details were acquired, and several similar models were built locally at CARDE's Propulsion Wing.[16] The PTV rocket design was kept relatively simple at first in order to facilitate the manufacture and reliability of more advanced models as CARDE gained experience. Consisting of three basic elements (a motor, the propellant, and the casing), the PTV personified simplicity. Bristol Aircraft (Western) in Winnipeg oversaw construction of the seventeen-inch casings, and Canadair supplied the nose cone and the tail fin stabilizers. This process of manufacture changed later, however, when the contract for the construction of advanced Black Brants was transferred to Canadian industry, and other companies, personnel, and locations also became involved.

Although all launchers were capable of carrying a payload, the purpose of the first two, the PTV and the Black Brant IIA, was simply to prove the design. Both rockets employed a highly reliable 15KS25000 series motor within a solid composite propellant casing. Although the original design relied on thick layers of polyurethane-mica to contain the hot propellant gases, subsequent versions incorporated an improved technology known as in situ moulded asbestos-phenolic mats.[17] Overall a very successful motor, the 15KS25000 generated fifteen seconds of twenty-five thousand pounds thrust in flight (hence its designation), making it capable of boosting up to 108.86 kilograms of payload to an altitude of 95.56 kilometres. The PTV proved so reliable that it malfunctioned only once during its first twenty-two static firing tests between 1959 and 1963.[18]

The solid rocket fuel CARDE employed was appropriately named CARDEPLEX. Based on a solid crystalline oxidizer and an organic polymerizable binder, rather than the usual nitrocellulose and nitroglycerine components, it was designed to meet specific casing requirements that the standard nitro-based propellants simply could not satisfy. The propellant manufacture itself was a two-step process. First, ammonium perchlorate oxidizer was ground, sifted, and then mixed with a smaller amount of another fuel based on carbon, hydrogen, oxygen, and nitrogen to create the crystalline oxidizer. Then, after blending and curing, the oxidizer was again blended with a polyurethane binder and poured into the PTV engine casing. The casing itself was internally coated with a heat barrier restrictor and mica-filled polyurethane bonding agent that applied well to the steel wall and to the propellant. Once enclosed, it was ready to be fired.

After several successful tests, the CARDEPLEX propellant formulation was finalized in October 1958, and by December, Cameron's teams had completed construction of the first few PTVs and were ready to begin testing the new vehicle as a whole.[19] The first static firing of the PTV took place in February 1959, and though the results were not perfect, they were very promising.[20] Corrections to the PTV design were made based on the results and more tests conducted throughout the spring and summer, including at different temperatures and environmental conditions. It was anticipated that, with a little effort, a real launch could be attempted at the Churchill Research Range (CRR) during the autumn of that year.

The CRR site was ideal for launching Black Brant rockets. Located at Fort Churchill, it was also a Canadian military base that served the Defence Research Northern Laboratory. In 1955 Canada and the United States concluded an agreement to build and maintain a joint facility there to test fire the American Nike Ajax missile system under extreme cold weather conditions.[21] The range was subsequently employed during the International Geophysical Year and remained under American control until the project ended on 31 December 1958. Throughout the IGY, the United States carried the entire operating cost of the installation ($7.3 million), as well as providing all eighty-six sounding rockets for international scientific experiments. Two of these rockets were designated specifically for Canadian use and were flown in November 1958 mounted with nose cones instrumented by CARDE.[22]

In 1959 the range and facilities were scheduled to revert to Canadian ownership (except for any equipment the United States chose to remove), but the US Department of Defense Research and Development Office expressed an interest in reopening the site for further missile and rocketry use. A new agreement was completed, after which the US Army tasked the White Sands Missile Range to put the facility back into operation. As a result, the US Army constructed several new buildings and modernized the launch facilities that would prove more than suitable for Canadian rocketry needs. As Canada had previously cooperated with the United States on the use of the CRR, little work was required to allow it to test and fire its own rockets from the launch pads employed by the US Army (and later the US Air Force).[23] This cooperation between American and Canadian militaries at Fort Churchill, which predated the space age, was set to last for several more years.

More importantly, the location of the CRR provided a huge natural safe zone of impact that was necessary for the American missile tests and was thus more than suitable for the operation of Black Brant rockets.[24] Fort Churchill as a whole also had the great scientific benefit of lying near the middle of the zone of maximum auroral activity, and this was augmented by its centralization on the geomagnetic pole. This gave both the civilian and the defence scientists an ideal location from which to conduct ionosphere-related studies that were crucial to the development of missile defence and space flight. From the CRR, Black Brants were tested in these conditions and later used to collect scientific data about the ionosphere.

An inexpensive inclined launch rail system for the PTV was designed in the latter months of 1957. It consisted of a simple elevating boom that, when horizontal, could have a PTV underslung from three guide rails ready to fire. The boom and PTV were then simply elevated to any desired launch angle between seventy and eighty-two degrees, and the forward braces were bolted to ground anchors. Once everything was stabilized and secured, the PTV was ready for launch.[25]

The first inclined launch rail was installed at the CRR in the fall of 1958, approximately 150 metres south of the American Nike-Cajun assembly building that was constructed during the IGY. This provided an almost due east trajectory over the water, which was necessary since maximum safety during launches was required. As well, should any problems occur during firing, the PTV would simply plunge into the bay and thus be unlikely to cause serious damage, an important consideration given that, as a solid-propellant rocket, it was non-recoverable once it was airborne.

The first group of four PTV rockets was scheduled for launch from the CRR in September-October 1959. Since the PTVs were not designed to carry payloads but simply to test the quality of the CARDEPLEX propellant, the first two were heavily ballasted to ensure that a large stability margin was maintained at Mach 6, whereas the second two were lightly ballasted to what was considered the minimum telemetry weight.[26] Telemetry was measured using a standard 30 by 30 pulse duration modulated–frequency modulated carrier system with twenty-eight active channels. Accelerometers and skin thermistors were added to supply additional data on the vehicle performance and the effects of flying at high Mach through the lower atmosphere. Finally, a radar beacon was installed in the nose cone to aid tracking of the PTV throughout its trajectory, to evaluate drag, and to verify its high-altitude performance.[27]

The first two flights of the PTV – now renamed the Black Brant I (vehicle Nos. CC601 and CC602) – provided the benchmark from which to proceed.[28] Because rocket CC601 had rolled a great deal during flight, its telemetry was not considered of value, but rocket CC602 provided much valuable data. As well, the two experiments flown aboard CC602 – a sodium and an infrared photometer – performed well beyond expectations.[29]

The Black Brant II design was essentially a mature version of the Black Brant I. The rocket was based on the requirement of attaining from a near vertical launch a minimum altitude of 228.6 kilometres with a payload of 68.03 kilograms. Its governing philosophy was that the structural design would be as light as possible without compromising efficiency. The Black Brant II was also slightly longer than its predecessor at approximately 8.5 metres, and the length of the nose cone was also extended to 2.1 metres. Whereas the Black Brant IIA employed the 15KS25000 motor, the Black Brant IIB was designed for a longer-burning 23KS20000 propulsion unit.

Operations and support for Canadian rocket launches were also beginning to take shape. With the entry of the Black Brant II into service, the DRB, CARDE, and the various agencies sponsoring the flights collaborated on the development of procedures for the checkout, ground handling, and launching of each rocket, ensuring satisfactory firings from the pad every time. CARDE and the Canadian military organized dedicated casings and transport vehicles for rocket parts, and requested specifically designed vehicles required to support any expansion of the existing program. Finally, launchers intended for use with the Black Brant I were modified to increase the maximum launch angle from seventy to eighty-five degrees, and an Aerobee rocket launcher was modified to accommodate Black Brants as well. Though none of the launchers had yet been augmented to facilitate year-round firings, these small increases greatly expanded the potential range of operations for Canada's new "rocketeers."

Black Brant IIA test flights began in 1961 and proved generally successful. The design consistently performed well in flight and was subsequently utilized in no less than fifty-five launches between 1961 and 1966. The main employer of the Black Brant IIA was the Canadian university research community, usually sponsored by the NRC, though a few were also used by the US Air Force (USAF) Cambridge Research Laboratory. The IIA model, the first workhorse of the Black Brant fleet, was retired only in the mid-1970s after several years of service. Its motor lived on, however, in both Black Brant IV and VA configurations.[30]

As the Black Brant IIB was designed to meet a requirement for improved altitude using the new 23KS20000 motor rather than as a scientific payload booster, its production was terminated after only four test flights in 1963. The lessons learned from this interim launcher were incorporated into an improved IIA model and the forthcoming Black Brant III.[31]

Early Programmatic Issues and Politics

The initial success of the CRPP, and the subsequent flights of the Black Brant I and II rockets, simply whetted the appetites of both the defence and civilian scientific communities for an expanded space research program. Yet, in the absence of an overarching national-level body responsible for *all* aspects of space research and development in Canada, the agencies that most sought to employ Black Brant rockets were only loosely organized to coordinate what was sure to become an increasingly expensive and complicated task.

Ottawa had not created anything similar to NASA to govern the development of its own space program. Instead, several separate organizations were simultaneously pursuing rocketry and space projects, with very limited cooperation between them. Realizing that a more official forum was required for the planning and execution of these programs, Donald C. Rose, a senior scientist with the National Research Council, called together a group of pioneering space scientists, engineers, and advocates from across the country to participate in forming an official space program committee. Although delving into high-profile subjects such as the formulation of a national space policy was beyond the initial scope of this gathering, it was intent on generating some type of common guidance with respect to the official development, employment, and future evolution of Canada's new rocket system.

The chosen group gathered unofficially at the NRC on 7 April 1959, where, among other preliminary business, it settled on a name for itself – the Associate Committee on Space Research (ACSR). It met for the first time as an official body on 2 October 1959.[32] Rose acted as chairman of the new twenty-member group, which included senior staff from a dozen universities, the NRC, the DRB, CARDE, and the Department of Transport. After calling the meeting to order, Rose introduced E.W.R. Steacie, the recently appointed president of the NRC, who welcomed everyone before briefly outlining the objectives of the group assembled before him.

Steacie announced plainly that "this committee will provide a mechanism for participation in rocket firings by the universities as well as government departments. It will also serve as the Canadian national committee for the International Council of Scientific Unions (ICSU) Special Committee on Space Research (COSPAR), and will ensure that Canadian representatives at the United Nations are kept informed of the views of Canadian scientists regarding activities in space."[33] However, though UN-related space activities were deemed important, the ACSR's record demonstrates that, during its early years, the issue of rocketry for science, technology, and defence lay at the centre of its efforts.

Though such a focus may at first appear misplaced or even politically naive, it made perfect sense in the context of the period. In 1959 space exploration was in its infancy, and even the two main adversaries, the US and the USSR, were still mastering the art of escaping Earth's gravity. Without the development of rockets, nothing – alive or otherwise – had any hope of attaining outer space. As well, rockets designed to reach the upper atmosphere or farther were simply ballistic missiles without a warhead. As such, they had huge potential, not only as a platform for exploration and science but also for weapons and defence. Thus, with an acrimonious relationship evolving between the two superpowers,

FIGURE 12 Brilliant and headstrong, Edgar William Richard Steacie, OBE, served with the NRC during the Second World War and as its president between 1952 and 1962. National Research Council Archives.

investment in rocketry development was considered not only scientifically informed but also militarily prudent.

Once the status of Canada's rocket-based research program was assessed, the ACSR meeting concentrated on the coordination of future activities. Whereas the NRC and the universities were primarily interested in expanding their overall research in space science, the defence sector was keenly interested in upper atmospheric research, atmospheric seeding experiments, atmospheric effects on the re-entry of objects, and larger-scale rocket studies and testing.[34] That both sectors could work cooperatively was feasible; it was just a matter of determining exactly how.

Many questions were not resolved at the meeting. For example, it was uncertain whether the NRC or the DRB should become responsible for the design and construction of rockets and payload experiments. As well, the details of how nose cones, with their delicate cargos, would be transported to the remote firing sites at the Churchill Research Range for checkout and launch had yet to be determined. The DRB and the Department of Defence Production (DDP) had initiated a series of studies on establishing some form of indigenous rocket supplier in Canada, perhaps in cooperation with industry, but at least a year would pass before a detailed report on the subject was ready for review. There were also concerns over present launch facilities and range control issues, and whether enough human resources were available to carry out sustained launching activities over a longer period of time. Again, unlike NASA, Canada was not actively engaged in developing a large workforce in space-related skills and trades. All of these issues were raised during the first ACSR meeting, immediately providing the members with a number of topics requiring their best efforts and attention.

Interestingly, the details of this meeting and the subjects discussed were treated as secret, with the chairman reminding all present that no one should communicate anything to the press or public until it was deemed appropriate to do so.[35] The order, not intended with severity, was rather a subtle reflection of the great sensitivity with which any issues related to the novelty of space exploration were treated during the late 1950s. With that comment, the first meeting of the Associate Committee on Space Research adjourned.

Though the creation of the ACSR certainly focused the wide range of interest in Canada's rocketry program and empowered it with political legitimacy, the ACSR executive committee was still faced with the difficult task of validating its plan to expand the current rocket program. Since the NRC and the universities had no way of initiating sustainable production, launch, and control facilities, another agency would have to provide them. The military already had some of these services, but hopes that the defence community would drive the growth in Canadian rocketry were quickly extinguished.

At a 10 December 1959 executive meeting of the ACSR, J.E. Keyston, deputy director of the Defence Research Board, explained why. As he put it, there was perhaps some hope that the Canadian forces would develop strategic missile systems similar to those of the United States. Canada was in the process of acquiring and equipping the nuclear-capable Honest John surface-to-surface missile for its land forces stationed in Europe; therefore, the idea that the RCAF would follow suit with its own system might have seemed plausible to the rocketry community. However, Cabinet had stressed to the Department of National Defence (DND) that the delivery systems currently within the Canadian forces arsenal were deemed more than adequate for its limited stock of high-yield ordnance, and thus Ottawa had no intention of developing large multi-stage strategic missiles for military use of any kind.[36]

Previous interpretations have identified this government decision as a step toward disarmament, but it was more simply a savvy assessment from a pragmatic Cabinet. Canada already had the platforms required to deliver nuclear weapons by other means and did not need to create a new and expensive strategic missile system similar to that of the United States. Canada could make its contribution to the collective defence of North America elsewhere without engaging in a costly large-scale launcher program.

This was an important if not confusing decision to some involved in Canadian space development. Although the government encouraged rocketry and missile systems advancement, it was setting clear limits on the level of capability it expected to achieve and sustain. With these restrictions, the DRB had no official mandate to retain expert knowledge in sophisticated launch systems to advise the military staff; instead, it was directed to maintain the small group of ballistics personnel it employed at CARDE to act as its rocketry advisers when needed. Personnel who had worked on the Velvet Glove project, and who were currently engaged with the Black Brant, were expected to suffice. If additional expertise were required at a later date, the RCAF could simply recall any of its personnel then on exchange with space and missile systems divisions in the United States.[37]

Mass production of rocketry was not even required for defence research. Since CARDE needed only a small number of research rockets every year for its activities, it had no necessity to build the additional production or control facilities that larger launch platforms demanded. Nor was Canada yet in the business of multiple payload launches, the other main impetus for a sustainable launching operation. Unfortunately for the ACSR, any justification for an expanded rocket program would have to come from Canadian scientific, not military, needs.[38]

Another suggestion was tabled at the December ACSR executive meeting. If the ACSR could provide Keyston with a submission stating the need for a Canadian rocket, and hence the production and control facilities that went with it, he would be in a better position to have the DRB request that additional resources be allocated to those already in place at the Churchill Research Range. Further, he was interested in knowing when a two-stage or larger rocket might be necessary, as this would influence the level of resources required. Opting to use American rockets instead was also discussed, but this immediately raised questions about availability, maintenance, fitting into existing handling facilities, and Canadian know-how.[39] Why these were such great concerns is not entirely explained in the meeting minutes, especially as the United States owned and operated the CRR during the period when the ACSR intended to execute its rocket research program there. In a sense, Canadian rocket teams were already depending on American facilities and technical assistance. Ultimately, the committee decided that, for the time being, the best way to proceed was with a program based on existing capabilities and resources. It was also agreed that American Nike-Cajun rockets, which were previously employed by Canadian scientists at Fort Churchill, would continue to be used for some future Canadian experiments until the DRB's own launcher capabilities evolved further.

Subsequent ACSR meetings in 1960 produced a tentative budget and agenda for the first series of Canadian scientific rocket flights. The number of Black Brants needed for initial university space science experiments was determined, as was the substantiation for the development of various multi-stage versions of the rocket for larger scientific payloads. If this activity were sustained or expanded over the next couple of years, it was hoped that larger multi-stage launchers such as the American Scout rocket might become a more viable option for Canadian use. CARDE was identified as the best agency to proceed with Canada's rocketry development, but it was acknowledged that commercial contractors might be required to support the stage of the program.[40] As well, it was determined that arrangements be initiated for ownership of the Churchill Research Range launching facilities to revert from the United States to the Canadian forces, with an expected handover some time in 1962 or 1963.[41]

Transition: Bristol Aerospace and Black Brant III and IV

With the Black Brant IIA and IIB projects well under way, the DRB turned its attention to planning future sounding rocket systems. The success of the first two rockets in the series had demonstrated that Canada clearly had the potential to build a more sophisticated family of sounding rockets for both civilian and defence scientific research. The US DOD and NASA had also expressed an early interest in Canada's Black Brant program. Although CARDE had assumed the lead in the initial project, it simply did not have the capability or responsibility to undertake large-scale production of Black Brant launchers. It recommended to the DRB that perhaps one of the civilian companies involved in the project could be made the prime contractor.

The Department of Defence Production conducted extensive market surveys in 1957 and 1958, establishing which companies would be best suited to take over the main portion of the Black Brant project. A major concern was whether Canadian industry would be able to provide adequate technical support for an indigenous launch program, but it found that Bristol Aircraft (Western) (BAL),

FIGURE 13 Canada's rocket pioneer. As an engineer with Bristol Aerospace in Winnipeg, contracted to the Defence Research Board, Albert Fia led the development of the country's first official launcher program during the height of the Cold War. Courtesy Magellan Aerospace Ltd.

one of CARDE's original rocket suppliers, could meet the basic requirements. Further wide-ranging market surveys conducted by both Bristol and the government in 1959 led to a solid proposal for a Canadian rocket industry in November. A year later, Ottawa formally awarded a contract to BAL for three new versions of the Black Brant, to be developed as a joint DRB-BAL project. With this decision, Canada's rocket industry was officially born.

Like CARDE, Bristol Aircraft was an obvious choice to undertake a pioneering role in Canada's rocket program. Headed by Stanley Haggett, the company had both the necessary technological base and the manufacturing capability. The engineers and scientists under the management of Murray Auld possessed all the specialist techniques and skills needed in working with high tensile and heat- and corrosion-resistant metals such as those employed in launcher systems.[42] Finally, Bristol's plant was also located conveniently close to the Churchill Research Range, where most Black Brants would be launched.

A young and talented engineer named Albert Fia headed the Black Brant technical team at Bristol. He held a degree in electronic and electrical engineering and had extensive experience in the development of missile systems for the Canadian Army. Fia was also a member of both the Manitoba and Ontario Associations of Professional Engineers and was considered an excellent leader by his peers. Joining Bristol in 1958, just as the launcher contracts were tendered to industry, he soon found himself assigned to head the newly created Special Projects Group that would design and build Black Brants for the DRB.[43] He lost no time in getting to grips with the challenge laid out before him.

Three new versions of the Black Brant to be designed by his Special Projects Group covered a range of altitudes and payload weights based on the needs of potential users and the environment to which the research instruments would be subjected. The market surveys of the DDP and BAL had revealed that scientists desired a minimum-cost highly reliable rocket able to carry payloads weighing between 4.5 and 136.0 kilos to heights of eighty to a thousand nautical miles, with little dispersion in performance.[44] Equally important, environmental conditions restricted the acceleration load on the instruments carried to forty Gs and the temperature to 125.0°F (51.6°C). Such demands were optimistic to say the least, and from the beginning, compromises were necessary if any success was to be achieved within these limits.

The first alteration to the design specifications was a reduction of the maximum desired altitude from a thousand to six hundred miles. Next, Fia's team simplified the development process by incorporating a number of existing CARDE components into the three new designs. It was decided that Bristol would continue to employ the Black Brant I's 15KS25000 motor and the DRB's CARDEPLEX propellant. Finally, Bristol introduced new steels with increased tensile strengths and improved ablative coatings that provided rigidity-to-weight ratios very close to the desired specifications. All in all, the Special Projects Group achieved remarkable success given the difficulties posed by the original parameters.[45]

The Black Brant III was a scaled-down version of the Black Brant II, using a ten-inch-diameter vehicle in place of its predecessor's seventeen-inch casing and CARDE's newly developed 9KS11000 rocket motor. Approximately nineteen feet long, the Black Brant III was capable of carrying a eighteen-kilogram payload to a height of 178 kilometres.[46] Testing of the new motor began in late 1961, with fifty-three static firings carried out at CARDE followed by another twenty structural and aerodynamic tests at Bristol.[47] By May 1962, everything on the Black Brant III had been tested, retested, and tested once more, leaving only the final exam for the rocket – proving that it could fly.

The original plan to launch the Black Brant III from the CRR had to be scrapped after a terrible fire devastated a good portion of the facility. Desperately, Bristol sought an alternative launch site and fortunately was able to secure a launch pad at the Wallops Island rocket range just off the Virginia coast. The entire CARDE-BAL team and four Black Brant IIIs were flown to the test facility aboard an RCAF C-130 Hercules transport plane. Ralph Bullock, an electronics engineer with Bristol, remembered that on arrival the pilot of the C-130 executed

FIGURE 14 Researchers and technicians handling a nose cone from a Black Brant rocket. These high-altitude experiments paved the way for much larger national research projects. Communications Research Centre, 60-RPL-0942.

a pre-landing show of aerobatics that he considered hardly appropriate for a transport plane, let alone his precious cargo. Fortunately, all arrived safely, and the team spent the next few weeks conducting final assembly and checkout of the four Black Brant IIIs.[48]

On 15 June 1962, after nearly four years of effort, the CARDE-BAL team launched the first two Black Brant III rockets. Carrying net payloads of 42.1 kilos and 43.1 kilos, respectively, the rockets performed impressively but less well than estimated. The first attained a maximum velocity of 1,712 metres per second and a range of 98 kilometres, whereas the second travelled slightly slower but farther at 1,706 metres per second to a range of 137.4 kilometres.[49] On-board instrumentation on both vehicles recorded a large but short-lived pitching disturbance six seconds after liftoff that diminished their peak altitudes by nearly 20 percent. The team realized that, if the Black Brant III were to fly properly, this flaw must be rectified.

Despite adjustments made by the CARDE-BAL team, the next two vehicles suffered similar problems. A third vehicle launched on 19 June 1962 had been fitted with additional fin cuffs to reinforce the stabilizers, but its lighter payload was wrenched so severely by the pitching disturbance seven seconds after liftoff that all telemetry with the vehicle was lost. Tracking films later revealed that it had righted itself afterward and continued to an estimated altitude of 125.5 kilometres, but the problem of stability remained unsolved.[50]

The fourth and final launch of the Black Brant IIIs brought to Wallops Island took place a week later on 28 June 1962. For this flight, the CARDE-BAL team attempted to keep the vehicle properly positioned by spinning it at three revolutions per second while it was in flight. It was hoped that this gyroscopic spin would produce sufficient stabilization to keep the last Black Brant III from suffering the fate of its predecessors. As well, the rocket was fitted with a heavier nozzle. The tail stabilizer fins were canted accordingly, and the rocket fired while the Canadians held their breath and the Americans looked on with reservation.

At first, things looked promising. The Canadian team let out a sigh of relief as the Black Brant III passed the point of previous instability without incident and continued to fly skyward. However, as it climbed it began a graceful but not desirable helical motion during which, as one observer described, it used up a great deal of sky before recovering itself and heading off in a completely new direction.[51] The rocket reached a mere twenty-four kilometres altitude before plummeting back to Earth. The initial tests of Canada's newest rocket were over.

The CARDE-BAL team returned to Canada in July to painstakingly pore over the telemetry and tracking film records of the four launches. There was much work to be done. The largest problem was obviously the stabilization of the rocket, and until it was remedied, production of the Black Brant could not proceed. In one of the corrections, the original fin stabilizer assembly was replaced with three stronger and lighter aluminum single-wedge fins employing a more pliable plastic insulation called Avcoat. This new material demonstrated tremendous qualities – a sheet only sixty thousandths of an inch thick was capable of reducing an external temperature of 538°C to a surface temperature of only 149°C. As well, the fibreglass wrap originally used on the outside of the motor and nose cone to keep them cool was replaced with an internal payload-insulating blanket. Design changes were also made to the motor case liner to give the rocket more thrust and a lighter nozzle. It was hoped that these modifications would prove successful.

Two redesigned rockets were taken to Wallops Island and successfully launched on 13 December 1962. Both achieved near perfect flights and returned full telemetry right to splashdown. Members of the launch team congratulated one another and returned home, this time more confident that the Black Brant III was ready for more general use. However, frustrations returned in July 1963

FIGURE 15 A series of Black Brant rockets, including the new Black Brant IIA model in the foreground, being positioned for launch at the Churchill Research Range, July 1963. Courtesy Magellan Aerospace Ltd.

when, in preparation for competing for a United States Navy contract, a Black Brant III fired from Point Mugu Test Range lost control shortly after takeoff, once again because of lateral disturbances. The telemetry and nose assembly were also lost, and although tracking films showed that the rocket recovered and continued its flight, no data were returned to the launch team.

The failed test of July did not reassure the team going into the American sounding rocket competition that winter. However, the launch on 7 November 1963 was a near perfect flight. Sporting a spin-balanced nose assembly, the Black Brant III outperformed its expectations, but the US Navy remained unconvinced of its reliability, and an order for Canadian sounding rockets never materialized. Demoralized yet not defeated, Albert Fia's Special Projects Group carried on with a final test launch the following spring. Fired from a newly reopened

FIGURE 16 A Black Brant IIIA lifts off from an all-weather-hardened launch shelter at the Churchill Research Range, April 1964. Courtesy Magellan Aerospace Ltd.

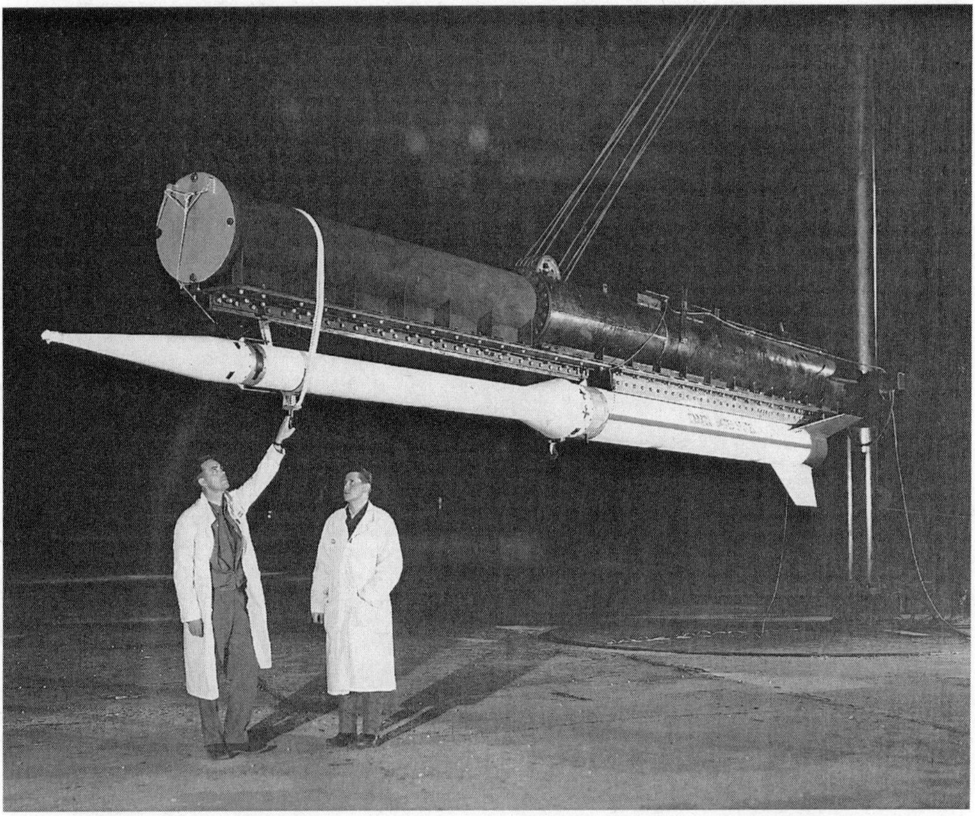

FIGURE 17 Canadian defence scientists prepare a Black Brant IV for launch at Wallops Island. Communications Research Centre, 68-16306.

Churchill Research Range, the last of the initial Black Brant III rockets was sent skyward on 21 April 1964. It performed well but nonetheless left the engineers with questions about their design concepts. Answers would be sought in the next test phase.

Unlike its predecessor, the Black Brant IVA was conceived as a two-stage rocket incorporating the Black Brant II and Black Brant IIIA motors. The new rocket was expected to carry up to eighteen kilos of payload to altitudes of 856 kilometres, much higher than the Black Brant III or its successor, the Black Brant V. It was also the first two-stage launcher attempt by Albert Fia's Special Projects Group, and he chose a close colleague, an engineer named Harry Sevier, to lead the effort.

Sevier assembled a ten-man team to tackle the Black Brant IV project. The point of utilizing the Black Brant II and IIIA motors was to marry these two

FIGURE 18 A DRB technician inspects second-stage sections for the Black Brant IV prior to assembly for launch. Courtesy Magellan Aerospace Ltd.

stages to produce a light yet robust launch vehicle. However, though the two separate stages were structurally resilient, the joint between them was another matter. The design and redesign of this part of the Black Brant IV would require several attempts before it worked properly.

The dramatically increased burnout altitude for the second stage of the Black Brant IV – approximately 30.5 kilometres – called for a longer and larger diameter

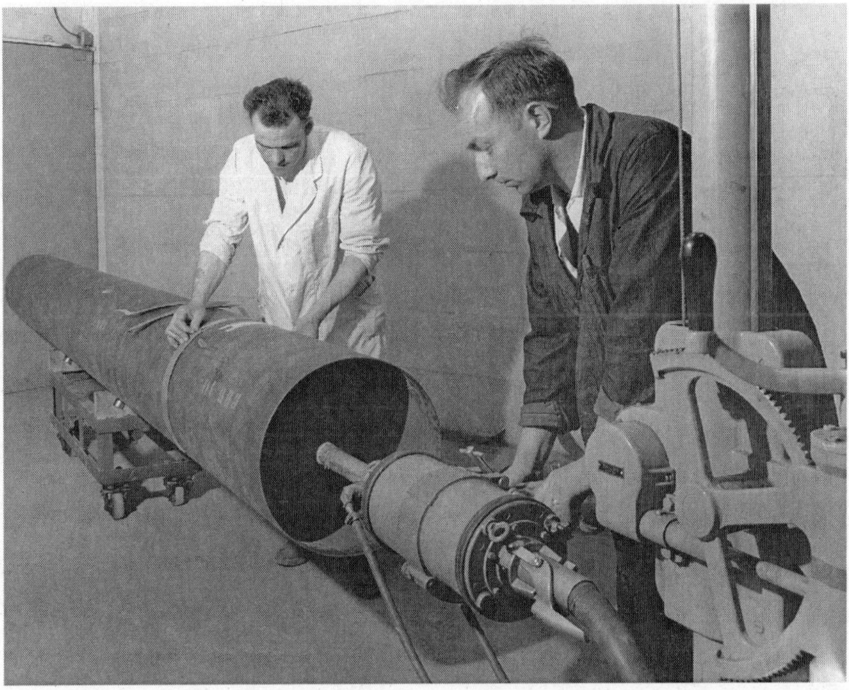

FIGURE 19 Bristol Aerospace technicians verify the circumference of a Black Brant IV first stage prior to filling with solid propellant. Courtesy Magellan Aerospace Ltd.

exit cone for the nozzle. Also, the upper stage required an odd-looking titanium conical device instead of fins, to keep the rocket stable in flight.[52] To facilitate separation, the two stages were not connected to each other: instead, the top portion rested on the bottom portion, using a sliding fit. A drag ring was mounted on the first, or "booster," stage so that when it stopped firing its higher drag would cause it to decelerate more rapidly than the upper stage and thus simply pull away. The goal was to achieve smooth separation, and the design appeared valid, at least on paper.

While the first Black Brant IVs were being readied for testing, preparations were made to collect telemetry from the flights. Though the newly rebuilt Churchill Research Range was capable of providing all the necessary diagnostics for Black Brant IV launches, the high altitudes expected from the rocket merited the use of backup trajectory recorders at the Prince Albert, Saskatchewan, tracking facility. Anxiously, the Bristol team members held their breath as Black Brant IVA vehicle No. 01 left the launch pad on 24 June 1964.

Despite their best efforts, the Bristol team faced another failure. The Black Brant IVA shot into the sky without difficulty, and the first stage appeared to be

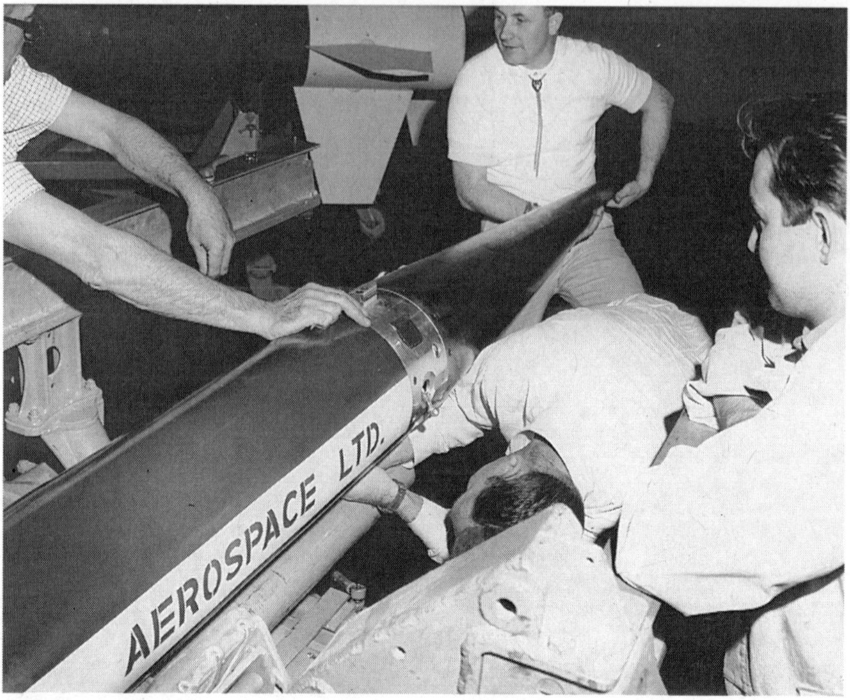

FIGURE 20 Tom Morgan supports Black Brant III payload while NRL technician Ed Wilder checks the experiment door. Jim Martin *(left)* and Churchill Range crew Bob Ruehlen look on, c. 1965. Courtesy Magellan Aerospace Ltd.

firing as planned. At thirteen seconds into the flight, when the second stage was supposed to separate, disaster struck. The ground team saw the exhaust trail wobble and then heard an explosion. Flight telemetry was suddenly lost. Still, the sustainer engine on the second stage ignited and carried on, but by this point the Black Brant IVA was so far off trajectory that it barely reached an apogee of 470 kilometres versus the anticipated 734.[53]

The next attempt was made on 2 July. Its fate resembled that of its predecessor, but embedded telemetry later revealed that it had staged too early and that, immediately prior to separation, the internal pressure was almost twice the outside ambient pressure.[54] The Bristol Aerospace official history adds further explanation, noting that the initial Black Brant IV flights had failed because of "inadequate inter-stage pressure venting, which in the absence of any structural joint between the stages, had prematurely pumped the rockets apart. As soon as the separation began, trapped inter-stage air escaped and the thrusting booster immediately drove up into the sustainer nozzle, producing violent oscillations – and collapse of the nose."[55]

FIGURE 21 Defence Research Telecommunications Establishment (DRTE) payload on a Black Brant IV rocket at NASA, Wallops Island, 1968. Courtesy Magellan Aerospace Ltd.

A few changes were made to improve the system – namely, the installation of a proper inter-stage venting process, a solid explosive bolt to connect the two stages, and flush-mounted booster drag flaps that would deploy when the inter-stage bolt was cut. In essence, the new design allowed the Black Brant IVA to separate only when permitted but to do so very quickly.[56]

The Black Brant IVA launch team returned to the Churchill Research Range in January 1965 with two revamped rockets to determine whether the upgrades were sufficient. Finally, after months of effort, Bristol tasted success. Vehicles No. 3 and No. 4 flew textbook flights, and two follow-on flights, also faultless, brought an agreeable conclusion to the Black Brant IVA project. An augmented version, the Black Brant IVB, went on to become a very successful sounding rocket used for many years by customers around the world.[57]

Final Success: Black Brant V

With each configuration, the scientists and engineers altered design parameters and introduced new motors and equipment with every intention of subjecting the rocket to rigorous testing that would push its aerodynamic limits. Although success was hoped for with each flight, it was perceived as unlikely. Yet, with each failure came valuable lessons that, when applied, ultimately resulted in the successful completion of the overall program.

The final phase of the DND-sponsored launcher project was the design and flight of the Black Brant V. The Black Brant VA and VB models differed little in their external appearance, both at seventeen inches in diameter, twenty-four feet long, and tailed with three stabilizing fins. To lighten the structure, thus increasing the range and altitude of the new rocket, the Black Brant VA exploited the mechanical interchangeability of the motor cases used for the 15KS25000 and the 26KS20000 engines. This allowed for a lighter casing, by using a lighter motor that could deliver almost twice the performance. The former engine was also employed on the VB model.[58]

However, the new motor design caused some initial problems: a failure during the first static test revealed inadequacies in the liner, which was required to insulate the highly stressed motor tube from the extreme temperature of the burning propellant. It took several redesigns of the engine and an additional twelve static firings before the DRB and Bristol engineers felt confident that the rocket and its follow-on VB model were both ready to fly.[59]

Other ways to lighten the structure were explored. The large and heavy magnesium tail fins normally fitted to Black Brant IIs and IIIs were replaced with smaller and lighter units consisting of a thinner aerofoil section that produced less supersonic drag. The fin itself was not the usual solid sheet of metal; instead, it was an aluminum honeycomb composite with bonded aluminum sheet skins. The whole construction was then coated with Avcoat. Many of the scientists were skeptical of the feasibility of this design, but from the outset the new fins worked as expected.[60]

Built concurrently with the Black Brant IV, the first low-capacity Black Brant VA test model flew on 16 April 1964, nearly three months before its sibling. It performed admirably and paved the way for tests of the VB. The first Black Brant VB launched from the CRR on 12 June 1965, carrying 140 kilos of instruments to approximately 378 kilometres altitude. The flight was flawless, easily demonstrating that the DRB-Bristol team had honed its skills to the point where few mistakes, if any, were made. The Black Brant VB continued to perform without error and was quickly adapted to carry scientific payloads even before testing was finished. On 16 August 1966, a Black Brant VB laden with a research payload from the Max Planck Institute of West Germany was lofted 391 kilometres above the earth where it released a cloud of barium. The resulting artificial aurora was visible as far away as Winnipeg and generated a boon of data for its researchers. The experiment was repeated three days later with another Black Brant VB, which in turn produced another textbook flight.[61]

The remaining Black Brant VB test flights were equally successful, bringing a satisfying conclusion to nearly ten years of design and testing. It was a bitter-sweet end to what many Black Brant engineers called "a labour of love."[62] When the last of the Black Brant Vs had flown, both the DRB and Bristol could look back with a sense of pride in what had been achieved. Essentially, in a little over a decade, scientists, soldiers, and engineers had given their country its very own rocket.

False Hopes for a Satellite Launch Program

As the testing of the Black Brant V drew to a close, discussion began regarding the next phase of Canada's rocketry program. Would the DRB and/or the NRC continue to support the development of larger-scale rockets, or would a commercial agency such as Bristol take the lead? Though it remained uncertain exactly who or what agency would direct future launch operations in Canada, few within the rocketry community seriously doubted that there would be a next phase. After the success of Canada's sounding rocket program and the launch of its first two satellites, it seemed only natural that the development of an all-Canadian satellite launch vehicle would come next.[63]

Proposals for future launch vehicles were plentiful. The advocates for larger rockets tabled a number of alternatives ranging from enlarged Black Brant models to mimicking American solid and liquid fuelled multi-staged launchers. Though many ideas were floated, three concepts were of particular interest to Canadian space scientists and engineers. The first two employed rockets, whereas the third examined the possibility of placing small satellites in orbit using a large-bore gun.

Of the two options focusing on rocketry, the first evolved from existing Black Brant technology, and the second involved the acquisition of the American Scout two-stage launch vehicle system and tailoring it to Canadian needs. Both concepts went through various stages of design and development with the combined support of limited government funding, the DRB, and university research. Success in bringing either to fruition was fleeting but not for lack of effort.

In 1967 scientists at the High Speed Aerodynamics Section, National Aeronautical Establishment of the NRC, presented a study on the employment of Black Brant VB motors as booster building blocks for the development of a full-scale Canadian satellite launcher.[64] The idea sensibly built on existing technology within the Black Brant program and established different configurations to construct a small series of three- and four-stage launch vehicles. One particular configuration, the 221X model, was considered by its advocates to be more efficient than the US Scout rocket, being able to lift 45.35 kilos into a 185 kilometre orbit. A flexible configuration, the 221X model was also designed so that it could even exchange its last solid propellant stage for a (yet to be developed) liquid-oxygen-hydrogen last stage, increasing its total lift capability to about 154.22 kilos.[65]

The following year, a scheme for the employment of the American-built Scout rocket was published in the *Canadian Aeronautics and Space Journal.*[66] Arguing that proposals for an all-Canadian system suffered, like all new programs, from a high risk of failure, unknown costs, and long domestic development schedules, the lobbyists for the Scout rocket option cited the precedent of employing American launchers and facilities (for Black Brants and Alouette-ISIS satellites) and suggested that an already tested American launcher would prove less expensive and more timely than a yet-to-be-created indigenous system. Government analysts agreed, stating in one of their official reports in 1967 that "it should be emphasized here that the development of a Canadian launcher of the Scout class is not an overly ambitious undertaking for a country which already is producing and launching multi-stage sounding rockets of the Black Brant type. The progression from sounding rockets to satellite launchers is fundamentally one of providing the necessary guidance and control to incline the flight path horizontally and insert a payload into orbit. The basic elements already exist of rocket motor technology, staging design, and launch and tracking complex at Churchill."[67]

The Scout was attractive for other reasons. Unlike most American launchers, it was specifically designed for non-military purposes and was thus obviously unsuitable as a weapons carrier.[68] This facilitated its export to other countries, such as Canada. Finally, the argument in its favour was further supported by the fact that strategic-level direction on the future of Canada's space program published in 1967 did not mention pursuing an independent launch system.[69] Given this, most observers in the Canadian space research community concluded that the Scout was the preferred option.

The four-stage solid fuel Scout stood roughly twenty-three metres tall and weighed 21,750 kilos fully loaded. A flight-proven design, it could launch, depending on the azimuth, anywhere from 210 to 270 kilos of payload into a 185-kilometre orbit. It also had the option of employing an upper stage designed for geo-transfer orbits, allowing payloads to be placed in much higher geosynchronous orbits as needed. Most importantly, perhaps, the Scout was small enough that it could be launched from the CRR without serious range tracking or safety concerns, allowing Canada to potentially place its own platforms into non-synchronous polar orbits.

The Scout proposal was revealed in tandem with another recommendation to build and launch an indigenous remote sensing orbital platform known as the Canadian earth resources evaluation satellite. This aimed at employing existing Canadian design capabilities to deliver a relatively inexpensive means of monitoring the country's natural resources from orbit. The project was considered a critical component in the development of new Canadian space applications, but because of the size and weight of the satellite, it could not be launched aboard anything smaller than a Scout rocket.[70]

Obviously, the second proposal could not fly without the first, and those involved in the two concepts worked hard to gain approval. In the end, however, the satellite project was never approved by government, and as all Canadian satellites were then being lofted by American rockets, there was some doubt by the late 1960s as to why the Scout was needed. Further degrading the Scout's applicability by that time was the fact that communications satellites, which were essentially the only type of satellite Canada planned to launch between 1969 and 1979, were too large for the Scout to lift and required much greater and more stable insertion orbits than the small four-stage rocket could achieve.[71]

Despite the fact that both ideas held tremendous technical merit and application, neither won political or financial support. In the late 1960s, many advocates within the DRB, the NRC, CARDE, and even industry promoted the creation of an indigenous Canadian launch capability, but this vision of the future was frustrated by the painful fact that, nearly a decade earlier, Cabinet had already concluded that embarking on such an endeavour was neither practical nor affordable. There was limited political support for the continued development of solid propellant rockets but none for initiating projects involving liquid fuelled or large-scale multi-stage launchers. In a memorandum prepared for the Department of External Affairs in 1959, DRB chairman A.H. Zimmerman noted, "since Canadian geography possesses no unique advantage in this area, there is no plan to undertake satellite launching." Later in the report, in a section titled "Facilities Required to Support Anticipated Canadian Program," Zimmerman added, "there is no intention to move toward Canadian manufacture or launching of very large multi-stage rockets to support any satellite program in which Canada would participate. Motors presently in use for this purpose by the United States are derived from the ICBM production effort of that nation. When it is considered that only because of the availability of the development and manufacturing skills of this large defence industry has the United States obtained its place in the space research field, the practicability of

Canadian manufacture of such rockets for space research alone is out of the question."[72]

Instead, the DRB (and subsequently the NRC) retained its near-term focus on the development of its current strengths in the rocketry field – specifically, design, test, instrumentation, telemetry, and electronics – while not concerning itself with any mid- or long-term planning unless the situation changed to merit a re-evaluation of that decision. Zimmerman may have chosen this rationale because all foreseeable satellite projects in Canada were likely to focus on communications, and though the Scout rocket could be used for molniya (polar) orbits, there was probably little requirement or opportunity to put such a complex system to other or greater use. A government assessment of Canadian rocketry interests in 1967 merely reinforced this position: "We do not consider that Canada should attempt at this time to provide satellite launch facilities to meet all program needs" but, as a result, it "will be necessary to purchase launches for communications satellites for at least the next decade."[73] Though not an ambitious or independent way forward for Canada's fledgling space program, this approach was nonetheless in line with other government decisions on strategic interests, and it served Canada's needs at the time.

The third available option was to use a high-velocity large-bore gun to launch small payload satellites into orbit. Although the concept of gun-launched rocketry was nothing new, no country had yet advanced the idea through to the design and test stage. Two projects to achieve this began in the latter half of the 1950s. The first was directed by Charles Murphy at the US Army's Ballistic Research Laboratory at Aberdeen, Maryland. The other project, under Gerald Bull, was initiated at the Aerophysics Wing at CARDE.[74]

Gerald Bull was a notably gifted Canadian scientist who was engaged in many aerodynamic and ballistics research projects at CARDE. When not working on larger-scale efforts, he often conducted many small, quick experiments during his own time, which were related to ballistics and particularly to high-velocity guns. It was during his tenure at CARDE that he developed his concept for gun-launched satellites.

Growing increasingly frustrated by the bureaucracy and the general lack of funding at CARDE for quick experiments, Bull resigned his position in early 1961. Fed up with what he often termed "cocktail scientist" officials and unable to advance his own research interests, he looked elsewhere for support. Interestingly, he refused several offers from American military organizations to work in the United States, preferring instead to remain in Canada with his young wife and family. Even Charles Murphy, who later became a life-long friend, was unable initially to lure Bull to come and work for the US military.

Opportunity in Canada soon presented itself. Shortly after Bull's departure from CARDE, Donald Mordell, the dean of engineering at McGill University, offered him a position on the faculty there. Bull gladly accepted, as a teaching job would offer him a stable income while at the same time providing opportunities to continue his research. Officially joining McGill University on 5 June 1961 at age thirty-four, Bull was one of the youngest people ever to be appointed to the faculty.

Over the summer of 1961, Bull and Mordell developed a plan called the High Altitude Research Program (HARP) to prove the gun-launch satellite concept through a series of test firings. Funding for HARP was originally sought from the DRB, but Bull had burnt many bridges and snubbed several ex-colleagues in his departure, thus ensuring that no financial help would be forthcoming from the DRB. Frustrated but not deterred, Mordell went to the Department of Defence Production (DDP) for assistance, and there his plan was received with greater enthusiasm. Convincing DDP officials of the merits of the many spinoffs that would come from the project, Mordell secured a verbal promise for a $500,000 grant. He then used this promise as collateral to secure an advance loan of $200,000 from McGill University's board of governors to get the project started right away. Bull also asked his friend and colleague Charles Murphy for further support, and the United States happily complied in order to secure direct access to the data and knowledge gained from the work.[75]

On the island of Barbados, using large-bore guns ranging from 12.70 to 43.18 centimetres, the HARP team successfully fired a number of fin-stabilized discarding sabot solid propellant rockets, named Martlet, to altitudes between 97 and 113 kilometres. On 19 November 1966, using a HARP gun designed by Bull, the US Army's Ballistics Research Lab fired an eighty-four-kilo Martlet to an altitude of 178.6 kilometres. It was, and remains, a world record for any fired projectile.[76]

Yet, despite the positive test results, the system suffered from several practical disadvantages. First, the payload had to be slender enough to fit into the gun-barrel tube, severely restricting the design of probes and satellites that could be launched in this way. Second, the Martlet left the tube discarding its sabot at a super-high velocity, at times accelerating to ten thousand times the force of gravity. This automatically ruled out attempting manned space flight from the mouth of a cannon. It also eliminated any possibility of carrying extremely sensitive payload instrumentation into orbit.[77]

Development of the Martlet proceeded under the HARP to the point where subsystems were tested and fired under high force of gravity conditions, but in the end it became apparent that the concept was not viable for launching

Canadian satellites. Sustained funding also remained a contentious issue throughout the life of the project, and Ottawa formally ended its support on 30 June 1967 in favour of pursuing other technologies.

After the termination of the HARP, Gerald Bull continued to conduct his research on large-bore cannons and munitions for two more decades through a variety of ventures and with the support of several governments. He developed long-range large-bore cannons for both South Africa and Iraq, and it is widely suspected that his association with the latter government led to his mysterious assassination outside his Brussels apartment on 22 March 1990. No one claimed credit for his murder, and Gerald Bull's killer was never found.[78]

Decision: Launch on Demand

With the final Black Brant V launch completed on 1 April 1967, the joint DRB-Bristol rocketry development program officially ended. Further versions of the Black Brant model were to follow, but these newer designs were not developed under DRB or CARDE guidance and control. Similarly, with the termination of the HARP, the "golden era" of defence-led rocketry in Canada was officially over, and non-defence-led rocketry projects also faced an uncertain future.

Additionally, the Canada-US cooperative defence research on launch ended on 1 January 1966, after which the Churchill Research Range was officially transferred back to complete Canadian control and subsequently placed under the management of the NRC and its Space Research Facilities Branch. The range continued to operate on an as-needed basis after that, funded jointly by the NRC and NASA, with the costs shared equally between the two agencies' respective governments. As well, Canada and the United States formed a Joint Range Policy Committee to oversee all future decisions related to firing specific rockets and payloads at the site. Each country nominated four representatives, one of whom was designated a co-chairman, to make up the eight-member committee. In 1966 and 1967, three of Canada's quartet came from the NRC (including a co-chairman) and one from the Department of Public Works. The American representation consisted of two members from NASA (including their co-chairman) and two from the US Air Force. Finally, the range itself continued to be operated by the American conglomerate Pan-American World Airways, with approximately 210 employees. The NRC also had eighteen personnel stationed at the CRR to support administration and flight operations.[79]

Overall, the Black Brant was a remarkably successful program that performed well beyond its original expectations. The Chapman Report, produced in 1967, noted that, during the initial DRB-Bristol development phase, 103 rockets had been successfully flown, most of them Black Brant IIA models.[80] There was no doubt that the Black Brant would go on to serve both Canadian and commercial interests abroad. The fact that it remains in use today is a testament to the quality of its original design. Even in 1966 and 1967, the sounding rocket market generated roughly $25 million per annum in the United States, and the Black Brant family of rockets rapidly gained wide acceptance as a reliable launch platform for scientific experiments in this market. The Black Brant generated $0.5 million in sales in 1966 alone, and it was later estimated that direct sales had reached $1.3 million by 1969. With the addition of supporting sales (extra user requirements, instrumentation, telemetry), the total amount was probably much higher. Similar government estimates put the total income for Black Brants and services from August 1966 to December 1967 at $3 million. By 1969 that amount had nearly doubled.[81]

However, financial success with the sounding rocket market did not automatically translate into a similar monetary success with larger launchers. Simply put, the larger the rocket, the more infrastructures and support it required. At least four or five launches a year were necessary if a launch facility were to pay for itself, but it seemed unlikely that Canada would ever have that many satellites in orbit, let alone during a single year. Essentially, the Canadian policy became launch on demand, and by the late 1960s, there was no demand from the Canadian military or defence science community. With defence out of the picture, only the Canadian universities and the NRC remained as potential customers, and their requirements could be met by the Black Brant. For larger civilian satellites, it was decided to purchase foreign launches as needed.

Yet Canada would not see the arrival of regular launch activity in the following decades. Although the US, the USSR, Britain, and France developed large-scale launchers as part of their strategic weapons programs and later for satellites or human space travel, Canada never had the same strategic imperative, especially after it abandoned its nuclear weapons arsenal, to take a similar path. With the decision made in the late 1950s not to engage in the design and construction of large-scale military rockets, Canada never evolved to the next level and thus never created the sophisticated platforms needed to carry larger satellites into orbit. Instead, it invested what it could into a modest program that served Canadian defence and scientific needs, and that, though not glamorous, was important, relevant, and ultimately successful. Such was the creation and evolution of rocketry in early Cold War Canada.

3

Defence and Discovery: Canada's Early Space Policies

CANADA ENTERED THE SPACE age both optimistic and uncertain about the role that rocketry and satellites would play in its future. National security dictated the evolution of many early space programs, and for Canada the situation was at first no different. Major projects such as the Churchill Research Range and the Black Brant, as well as the Alouette and the RCAF Space Defence Program (both discussed below), were all born out of Canada's many defence priorities. However, though these projects shared obvious defence science and technology applications, they evolved in relative isolation and were not considered the equal parts of a collective national strategy for long-term rocketry and space development. As will be seen later in this study, a lack of cohesion became glaringly self-evident when both Cabinet and DND were unable and unwilling to sustain any large-scale defence space programs beyond the late 1960s.

Despite several proposals to coordinate all of Canada's space activities under a single champion or organization during the 1960s, a general nescience among senior political leadership as to the potential long-term impact of space exploration and technology exploitation, and the distraction of more pressing domestic policy and government restructuring issues, resulted in little top-level direction for a formalized plan. The situation was further complicated by competing agendas as well as political conflict between Canada's scientific, technological, defence, and government communities during the 1950s and 1960s, with the result that the rocketry and space program advanced, not under a collective umbrella as some hoped, but in a piecemeal fashion that had serious ramifications for its future.

New Faces, New Directions

As the space age neared, many of those in government, research, and defence who had guided the country through its first decade of post-war technical modernization either left or retired from their posts. At the highest level, Lester Pearson's Liberal Party lost the 1957 federal election to the Progressive Conservatives under John George Diefenbaker. The landslide victory of 208 Conservative seats over the Liberals' 49 seats was the largest majority ever recorded to date in Canadian history.

Diefenbaker's government had a considerable impact on the formative years of Canadian space activities. His party assumed office just as the space race began, and obviously much of Canada's early space development depended on the health of its relationship with the United States. However, Diefenbaker came to power on a publicly perceived anti-American platform and at first seemed little interested in outer space or space cooperation with America unless he could draw good personal publicity from such issues. For example, it was common knowledge among his own staff that Diefenbaker routinely deviated from the facts in his speeches when discussing space program plans. Often ignorant of the scale of American space programs, he would announce similar Canadian scientific and technological initiatives without substance or understanding of the details or costs to "make a splash" with his audience.[1]

Limited in his knowledge of science and international relations (many members of his Cabinet had little or no knowledge of the subject at all), Diefenbaker nevertheless assumed personal control of the foreign affairs portfolio shortly after wining the election. He apparently wished to control matters and launch a new agenda rather than leave it to the existing civil service, which he greatly distrusted and derisively called "Pearsonalities," a reference to their supposed allegiance to the new Official Opposition.[2] The result was disastrous for the early evolution of Canadian space policy, never mind the development of new programs and projects that greatly depended on informed decision making to proceed.

In late 1957, perhaps suddenly appreciating the magnitude of the portfolio, Diefenbaker chose Sydney Smith, president of the University of Toronto, to take over external affairs. Sadly, Smith died unexpectedly soon after entering office, and Diefenbaker once again had to assume control of the portfolio. He subsequently replaced himself in 1959 with Howard Green, his minister of public works. A veteran politician and confidant of Diefenbaker but by no means adept at foreign affairs, Green understood little about space, science, and technology beyond the possibility that they might become a future battleground between the superpowers. Thus, he encouraged initiatives that advocated more international control of outer space, without fully understanding what was required or how such circumstances might influence Canada's own future interests.

Notwithstanding the considerable progress within the scientific community, the political importance of space and technological development in Canada remained unappreciated by Diefenbaker's government. Diefenbaker himself rejected informal proposals for the appointment of a science adviser and remained opposed to enlarging the Prime Minister's Office with any bureaucratic advisers, scientific or otherwise.[3] He seemed little interested in supporting technological development and personally cancelled a number of high-profile national efforts early in his tenure, including the CF-105 Arrow fighter interceptor previously slated for termination by the Liberals, as well as the High Energy Project. The general consensus was that Diefenbaker made science and technology a low priority.[4]

Still, by 1961 federal expenditures on scientific activity exceeded $220 million, almost seven times the amount invested on science in Canada during 1945. Nearly eighteen thousand Canadians worked in professional scientific and engineering organizations, which had also multiplied. No longer was national science and technology trapped within the traditional triumvirate of the Department of Agriculture, Mines and Technical Surveys, and the NRC. Now, government science included Forestry, Fisheries, Northern Affairs and National Resources, National Health and Welfare, Transport, and Veterans' Affairs. As well, the NRC had spawned a number of lesser agencies such as Atomic Energy of Canada Limited and the Medical Research Council.[5] Even then, however, this assembly of new scientifically oriented establishments was represented within Cabinet by two small groups only, one being the Privy Council Committee on Scientific and Industrial Research and the other the Advisory Panel for Scientific Policy. Neither was particularly effective or, it appears, taken very seriously by the senior leadership.

Keeping track of science and technology developments within government ultimately fell on Norman Robertson, a veteran diplomat recently appointed back to Ottawa as under-secretary of state for external affairs. Having previously served as ambassador to the United States, Robertson had some experience in dealing with missile, rocketry, and space issues, but without any ministerial oversight within Cabinet, he relied heavily on the heads of the NRC and the DRB for advice on the subject. His attention to every aspect of Canadian statecraft ensured that outer space activity would not be entirely ignored, but little more than passing notice could be expected, given that it was not a high priority with Diefenbaker's Cabinet.

The confused coordination of Canada's response to the 1962 Cuban Missile Crisis and the ensuing internal debate over nuclear weapons seriously damaged and fragmented Diefenbaker's caucus, and the Conservatives lost their bid for re-election in 1963. Pearson's Liberal Party returned to power after a hard-fought election with enough seats to form a minority government. Though certainly better prepared and equipped to handle nearly all aspects of Canada's international affairs, including those related to science and technology, neither Pearson nor his Cabinet seemed particularly interested in developing a national space agenda or policy while in government.[6] As for Pearson's personal reflections on the importance of the matter, insight may be drawn from the fact that he made not a single mention of outer space in any of his published memoirs.

After 1963 the Pearson government turned away from international issues dealing with science and technology, as internal government reorganization and professionalization of the civil service became top priorities. As political scientist G.B. Doern notes, "Pearson's views of science policy tended to be characterized by a genuine, but superficial, belief that science had to be given greater structural recognition in the inner circles of decision making. Pearson's ultimate agreement to create a Science Secretariat and a Science Council seems to have been the product of internal advice ... rather than any indigenous initiative developed by Pearson and the Liberals in their opposition days." Even under new leadership, science policy, and thus space policy, was left to the subordinates of government to guide and develop.

The first director of the Science Secretariat, Frank A. Forward, rarely met with the prime minister and had little influence in shaping national science programs. In fact, from his appointment on 30 April 1964 until mid-May 1965, he alone comprised the entire membership of the new secretariat. Though he later received three deputy directors, an executive secretary, and a small professional staff, all of his efforts during the next year were focused on special studies, legislative studies, and reviews of science policy in other countries rather than on advising the prime minister regarding a national scientific strategy for Canada. "The attitude of Pearson, to both the place of science and the need for advisors," observes Doern, "seems to have been one of general sympathy and benevolent encouragement, without much of a disposition for the machinery itself."[7] Given that Doern's analysis was published in 1972, soon after these events took place, his direct and confidential access to many of those involved in such decision-making processes should be considered authoritative in the absence of other studies.

Pearson's apparent lack of interest in Forward and his secretariat was soon felt. On 12 May 1966, Canada's first science adviser was informed that his organization was being permanently dissolved, replaced in government affairs by the newly created Science Council of Canada. Forward left his briefly held post soon after the announcement, having made very little difference in shaping the national space agenda.

Pearson remained in power until his retirement in 1968, but during his tenure Canada's space policy did not congeal as many had hoped it would. The absence of a clearly defined mandate or ministerial advocate during both Diefenbaker's and Pearson's terms resulted in a disjointed approach to both research and development as the formation of policy and programs was left up to the discretion and competing agendas of the agencies actively engaged in developing space projects. Yet even there, those who had chosen the original course were moving on, and those who replaced them brought their own ideas regarding Canada's space efforts.

Both the National Research Council and the Defence Research Board also came under new leadership during this period. E.W.R. Steacie, director of the Division of Chemistry, succeeded C.J. Mackenzie, the wartime head of the NRC in 1952. Unlike Mackenzie, the defence science guru who had close friends in Cabinet and considerable influence with the government, Steacie was generally distrustful of government involvement in scientific affairs and was not afraid to accuse the politicians of trying to direct a field in which he felt they had little understanding or right to meddle. Under his brief appointment, the NRC was often at odds with government on issues dealing with national research and development.

As mentioned, Omond Solandt, who had founded the DRB and shaped it during its first decade, retired from his post as DRB chairman in 1955 to be succeeded by A.H. Zimmerman, the board's vice-chairman. Zimmerman was not a charter member of the DRB; he had joined it in 1951 as the Department of Defence Production representative. Made DRB vice-chairman in 1955, he officially took over during the following year. Also, whereas Solandt was a pure research scientist, Zimmerman was an engineer, and his perceptions of science and technology were formed by his return to business after the war, not by national-level programs in theoretical research and development. The result was a pragmatic but at times short-sighted approach to running the DRB, favouring immediate and predictable returns from programs such as Alouette rather than long-term and perhaps riskier research and development goals such as an indigenous launch capability.

Reaction to Sputnik

Canadians were no less surprised than the Americans when, on the morning of 5 October 1957, national newspapers informed them that the USSR had successfully launched the first man-made object into space. For readers of the *Toronto Globe and Mail* in particular, the front page was filled with irony. Just below the Sputnik article was a story and photo detailing the rollout of the first AVRO CF-105 Arrow fighter interceptor jet. On the front page of this major Canadian newspaper, the advent of one technological achievement foreshadowed the demise of the other.

Within government, however, reactions by Canadian and American leaders seem incredibly different. Though senior American politicians and advisers were not necessarily surprised that a satellite had gone into orbit (recall that the United States had planned to launch a satellite as part of the IGY), they were considerably impressed by the technological magnitude of Sputnik. Furthermore, the success of Sputnik suggested an ominous capability to launch nuclear weapons to greater distances than ever before or even to put them into orbit.

In response to Sputnik, the United States brought a science advisory capability directly into the White House itself so that President Eisenhower could have immediate consultation on matters related to science, technology, and defence. The first special assistant to the president for science and technology was James Rhyne Killian, a prominent academic from Boston's Massachusetts Institute of Technology. As Killian later remarked in his published memoirs, "that a satellite had gone into orbit really did not surprise me ... The real significance of the news for me lay in two words: 'Russian' and '184 pounds.'"[8] The first American satellite, Vanguard, weighed in at about three and a half pounds, a rather pale achievement in comparison.

By contrast, little suggests that Ottawa was equally concerned about the advent of Sputnik. Neither Pearson nor Diefenbaker mentioned it in their personal memoirs of the period.[9] As well, Cabinet records reveal no significant decisions or statements regarding the Soviet launch, and there is no serious mention of outer space in the House of Commons debates until the following year.[10] Diefenbaker appointed no additional specialized scientific counsel to his staff, although arguably he already had the Privy Council Committee on Scientific and Industrial Research, the Advisory Panel for Scientific Policy, and the senior leadership of the NRC and the DRB at his disposal. Similarly, the government did not initially indicate any plans for an official response to Sputnik. This aloofness regarding such a monumental historical event is curious and lacks simple explanation.

Yet not all parties shared Ottawa's lack of interest. Canada's defence research community felt a real concern over the implications of recent Soviet missile and space achievements because the scientists and engineers knew what their adversaries had truly accomplished. "The announcement of the first flight testing of the Russian ICBM on 26 August, followed on 4 October by the launching of the first man-made satellite," noted one DRB scientist, "has not only pointed up the illusion of believing that the West has a well-established technical superiority, but in fact stresses the urgency of developing a thoroughly realistic approach to all of the complex problems of the next six, eight, or ten year period, in as short a time as possible."[11] The DRB was also concerned about keeping pace with Soviet resources. "If the USSR maintains her output of scientists and technicians at the present rate, the race will be lost to the West in point of numbers," the same report stated. "Our hope must therefore lie in conservation of effort and concentration on high quality."[12]

However, quality came at a price, and the DRB was increasingly struggling to meet the large debts incurred in the research and development of high technology. The ongoing cuts in Canada's defence spending toward the 1960s in the face of increasing wage, construction, and equipment costs was hard felt at the Defence Research Board. In the financial year 1956-57, approved salary increases alone cost the DRB $1.3 million in funds originally allocated for research. The additional costs were not covered by DND; instead, they were met by deferring all new construction, including a much-needed wind tunnel, reducing contracts with industry, and restricting the purchase of laboratory equipment required for proposed new programs, including space research.

Nevertheless, the importance of missile and space research did not entirely escape senior-level planners at DND. The protection of strategic air command bases in North America by means of integrated Canada-US air defence plans remained a priority, and this would eventually include anti-ICBM solutions. Similarly, strategic surveillance and reconnaissance were essential to early warning, intelligence analysis, and force protection, a capability that was rapidly transitioning toward space-based platforms. If, as one scenario suggested, a Soviet attack on the United States would be met in Canadian airspace, DND would need to defend against missile and space-based threats. Yet, before it could proceed with planning and operations, it required clear policy guidance from Cabinet and the Privy Council on what Canada's missile and space priorities would be. Was DND expected to prepare for an imminent Soviet attack, or was it reasonable to assume that the United States would protect Canada and the Canadian assets and facilities essential to its own survival? Defining strategic guidance was the first challenge to the evolution of Canada's defence and civilian space programs.

Attempting Internationalism

From the outset, Canada's own missile and space policy options were limited by politics and the size of the national economy. Unwilling and unable to keep technological pace with the rapidly expanding agendas of either the USSR or the United States, Canadian decision makers sought instead to build a program through niche capability, leveraging collaboration with the Americans and increasing Canada's influence in international space cooperation and control by attempting to champion third-party interests at international forums such as the United Nations. Some of these efforts brought success, whereas others did not, but all in some way influenced the first decade of Canada's national space policy and agenda.

When Cabinet and the Privy Council first considered a national space agenda in the summer of 1958, among the leading figures examining the issue was

Douglas V. LePan, a veteran of the Italian campaign, noted poet and writer, and also one of Robertson's assistant under-secretaries. Realizing that the superpowers were preparing to dominate space both militarily and commercially, he was concerned that, lacking similar technological capabilities or resources, Canada might soon find its own space interests restricted or even threatened. Worse, those who had space access would no doubt dictate the law and rules of space exploitation, which those who did not would be forced to accept. Still, unsure of the current status or potential capability of Canada's rocketry and space program, LePan sought out expert advice at both the NRC and the DRB while developing his agenda.

Consulting a wide range of administrators, scientists, and engineers from May through August 1958, LePan concluded that the best way for Canada to gain influence was to promote and, if possible, codify an agreement that made space control an international responsibility.[13] This way, third parties such as Canada could secure guaranteed space access and possibly influence how space would be used by all nations. The obvious venue to suggest such an option was the United Nations, and LePan immediately set out to prepare his argument for review by the secretary of state for external affairs and, later, the prime minister.

LePan's plan looked simple enough. Hoping for the creation of a peaceful space environment where Canadian interests would be secure and the country could prosper, LePan suggested that, first, the international community should declare space a sanctuary and, second, that Canada could play a lead role by building an international space flight development station where scientists and engineers from all nations could converge and share their research. Further, he suggested to the secretary of state that the Churchill Research Range, then still under combined Canadian-American military control, could be transformed into the proposed station.

Despite LePan's intelligence and generally good diplomatic knowledge, he was seriously uninformed about evolving rocketry and space events, with the result that his plan was ill-conceived. It placed too much weight on the notion that the lack of Van Allen belt effects, specifically radiation, on the Churchill Research Range because of its northerly location would make it a preferred launch facility for all future space flights. As well, LePan underestimated the acrimonious attitude of the Soviet Union toward any space cooperation, while overestimating the amount of influence that Canada could have in forcing the two superpowers to relinquish their obvious advantages in controlling access to outer space. He does not appear to have fully appreciated the larger political factors surrounding East-West Cold War relations, either preferring to ignore the *realpolitik* at the United Nations or implicitly trusting his various sources. Whichever the case, he received bad advice and had a poor appreciation of the geopolitical situation.

In correspondence to the secretary of state for external affairs, LePan later admitted that he had based a large part of his proposal upon informal conversations with J.E. Keyston, deputy director of the DRB, and an anonymous paper previously prepared by a group of concerned DRB scientists who worried about the potential weaponization of space.[14] Formal consultation was not slated to occur until the prime minister approved LePan's initial plan. This approach may have seemed politically savvy, but it was not well informed. Although LePan acknowledged that transferring the Churchill Research Range from American and Canadian military control might entail some difficulties, he felt that it should and could be done. Oddly, his suggestion virtually coincided with Cabinet approval of the renewal and expansion of the Canadian-American joint test facilities at Fort Churchill until 1962. As well, DND and the RCAF were in the process of formalizing a military space agenda, largely at the approval of the minister of national defence through the prime minister. The obvious discrepancy between the strategies being considered in the Department of External Affairs and what Cabinet was actually doing is both interesting and demonstrative of the disjointed – some have even labelled it schizophrenic – nature of Diefenbaker's government during a period of increasing East-West political tensions.

Although it was common knowledge that attempting cooperation with the Soviet Union was unlikely to bear fruit, LePan simultaneously forwarded a copy of his proposal to both the secretary of state and A.G. Campbell, a senior diplomat on the Canadian delegation at the United Nations. On 12 September, Campbell wired back a reply, indicating that all the Canadian delegation could hope to achieve was to provide for a committee to study the problem and to ensure that any future UN space cooperation committee included a Canadian representative.[15] Campbell then put off the issue until after Christmas.

Though Campbell had sidelined the issue for the moment, it remained high on LePan's agenda and was not ignored at home. On 17 December, in response to Cabinet's request for studies on Canada's current and proposed space program, Zimmerman submitted two papers to the secretary of the Privy Council Committee on Scientific and Industrial Research.[16] Both were titled "Space Science and Space Technology – A Summary of Points Affecting Canada's Future Position." The first was an executive summary, and the second provided a more detailed yet non-technical analysis of potential space options. Both also recounted the natural advantage of Canada's geography in contributing to the evolution of space science and highlighted the obvious benefits that space assets could provide to Canadian defence.

The reports advocated the establishment of a national space policy and an official organization or agency to administer Canada's growing space activities. Though they focused on expanding Canada's military space capability, they constituted the first request to the Privy Council Office to formulate an official space policy for the country. At the time, however, Cabinet was still evaluating its options and decided to retain the DRB as the national-level space advocate at least until LePan's proposal on international space control had run its course.

Perhaps over-optimistic regarding its diplomatic influence at the UN in the wake of its resolution of the Suez Crisis, Canada fell into the bitter quagmire of international space politics in early 1959. At the beginning of the 1960s, no formal or legal definition applied to space, and few rules governed how it would be used. Both superpowers had quickly realized its strategic importance, and neither was inclined to encourage legal boundaries in it, especially if this meant the possible forfeiture of its own unobstructed access to exploit space militarily and commercially. Although both superpowers publicly called for the peaceful use of space, neither was truly interested in making it a sanctuary. National security through outer space was too valuable to be left in the hands of the United Nations Security Council.

Equally important, the nature of space exploration represented serious challenges to the possibility of controlling it internationally. For example, space technology had a duality of purpose that made legal definition difficult. The United States employed its Redstone and Atlas launchers both as ICBMs and as boosters for its manned spaceflight program. The Soviet R-7 rocket was both an ICBM and the launcher that put cosmonaut Yuri Gagarin into orbit. Satellites had similar dual purposes. If either nation agreed to use space solely for peaceful ends, how could it justifiably build a launcher with the capability of sending men to the moon? Could not the same launcher lift a weapon of mass destruction into orbit? Who would adjudicate over space and technical programs? What international laws would apply? Ultimately, as similarly witnessed in air law, the duality of the technology meant that no one could realistically deny that space would eventually be used for defence purposes, making Canada's call for international control of space more altruistic than realistic.

Even senior Canadian diplomats were pessimistic about the LePan proposal. On 3 March 1959, correspondence between A.G. Campbell and John Holmes, another Pearson disciple and a member of the Canadian delegation at the UN, confirmed that drawing a sharp line between purely military or civilian use of space or related technologies would be difficult, if not impossible. More importantly, multilateralization of space was neither necessary nor expected, and such a move would do little to benefit the Western powers. Holmes also commented that any UN program promoting international space control would not be favourable if it restricted Canadian options, and any UN-controlled committee would almost certainly do just that.[17] Given this, one is left to wonder why Ottawa was interested in endorsing LePan's initiative at all.

However popular the international control of space may have been with Prime Minister Diefenbaker's Cabinet, Canada's UN delegation argued that it was simply not viable. Although A.G. Campbell reported back to Ottawa that Canada could champion a third-party space fraternity "to give reality to the claim to equal rights in outer space and to gain a position of influence for negotiating international control of space to ensure that it is used for only peaceful and scientific purposes," he conceded that the generally negative atmosphere within the UN regarding the initiative would probably kill it.[18] At the Department of External Affairs, Norman Robertson confirmed this in a memorandum prepared for the secretary of state the same week in which he detailed the status of space activities and the announced future plans for many countries, including the United States. Nothing suggested that either the United States or the Soviet Union would endorse international space control at the expense of its own national security or prestige.

Other options were needed. In Ottawa, Robertson recommended that instead of pushing for international legislation, Canada should focus on the development of its own agenda, which could then evolve through international forums. He suggested that the secretary of state support an upcoming DRB proposal for the creation of a Canadian space program, if for no other reason than to put Canada in a good position as junior partner to the United States and to gain access to the technological and economic benefits that would arise from the anticipated large-scale American program. Regardless of what objectives might be gleaned from the promotion of international space control, benefit to Canada clearly stood out as the primary aim of Robertson's proposal.

Robertson's approach was calculating if not shrewd. Championing international space control was seen as a way to "enhance Canada's prestige and status in the international community" while "foster[ing] Canadian scientific progress and provid[ing] a focus of interest to maintain and attract scientific and technical manpower."[19] Certainly, there was an interest in the promotion of outer space as a sanctuary but not if it jeopardized Canada's own agenda or its relationship with the US space program.

The proposed way ahead was not to match the US missile, rocketry, and outer space efforts but rather to cooperate with America and see what could be gained from the relationship. Far from encouraging altruistic multilateralism, the Department of External Affairs sought forums in which Canada might obtain an advantage or, better still, exclusive access to the resources of its closest allies.

In a confidential memorandum to the secretary of state commenting on American plans for space, Robertson noted that

> One point which appears to be relevant to the consideration of a Canadian programme is that, if Canadian industry is to have a reasonable chance at securing a share in the benefits of the expected United States [space] effort, it behooves us to ensure that government agencies are enabled to keep abreast of the course of developments so that Canada may claim status as a junior partner, at least in the area of space science, with an eye to the possibility of reaching ultimately a production-sharing agreement in the space technology area. Another point is that space exploration seems likely to prove one of the most prolific sources of stimuli to new ventures in many scientific disciplines and accordingly a suitable programme should serve to strengthen science generally in Canada.[20]

Ultimately, a middle road was chosen. The prime minister consented to the pursuit of a Canadian initiative for the international control of space so long as Canada could strongly influence the process, and he ordered both LePan and Campbell to proceed.[21] Consequently, over the summer, directives were issued to both the DRB and the NRC to commence studies on the technical feasibility of a Canadian initiative to improve the outlook for international cooperation in space research and peaceful uses.[22] On 5 June, the DRB nominated J.E. Keyston as its representative at the working group, with W.M. Cameron, the DRB's director of plans, as his alternate. On 10 June, the NRC replied with the nomination of D.C. Rose and J.D. Babbitt to the working group. All parties were asked to examine LePan's proposal, improve upon it, and return a feasibility report to Cabinet for consideration no later than the end of August.

The working group met during June and July. Neither A.H. Zimmerman nor E.W.R. Steacie were impressed with the Canadian initiative for international space control, especially with LePan's idea of transforming the Churchill Research Range into an international space flight development station.[23] Keyston felt that neither DND nor the DOD would give up their facilities at the Churchill Research Range, which at the time was proving invaluable to the research and development of anti-ballistic missile systems and defences for North America.

For his part, E.W.R. Steacie was less diplomatic about the whole idea. Upon entering Robertson's office in June 1959 for a private meeting with the undersecretary and Zimmerman, Steacie produced a comic book and threw it on the desk. It depicted space cadets, and Steacie suggested that it had much in common with the department's plan for the international control of space.[24] Retaining his official composure, Robertson listened as Steacie laid out several arguments against the proposal. Robertson later stated that Steacie feared that the "Prime Minister evidently intended to do something which was essentially silly scientifically in order to get a good speech in the United Nations," noting that the UN was not much interested in space activities and could offer little beyond insight into its various legal problems.

Steacie was also very displeased by the notion that politicians, not scientists, were determining Canada's national science agenda. "Evidently [Steacie felt] the Prime Minister's policy was not to ask whether or not an idea was any good but to tell the scientists what was going to take place," Robertson remarked in his report.[25] Although Zimmerman cited favourable examples of international science cooperation, Steacie argued vehemently that international laboratories never worked. Instead, he proposed that the funds be spent on improving Canadian laboratories and then inviting other countries to come and conduct space science research in support of Canadian interests. Though still not entirely sure of the real reasoning behind Steacie's malice toward fostering international co-operation, Robertson took the suggestion into consideration.[26]

Zimmerman then attempted to provide greater context to their concerns. He elaborated that many difficulties and rumours had arisen because the prime minister had deviated wildly from the planned message of a recently delivered speech on the future development of Canada's space program at the Prince Albert Radar Laboratory. Zimmerman stated that, before Canada created an establishment for international use, Ottawa should determine who would be interested in such a facility. He pointed out that most if not all of the current or likely spacefaring nations already had their own ranges, and since the NATO countries had been invited to work out bilateral agreements on the pattern of the DRB-NASA project, it was doubtful that the Churchill Research Range would be attractive to countries with a developed scientific capability.[27]

Staffing and financing posed problems as well. Both the DRB and the NRC were fully engaged with current projects and could not support any new ventures without jeopardizing existing Black Brant, Alouette, and other ongoing space science endeavours. Additionally, the United States alone had already invested $14 million in the Churchill Research Range site, and both American and Canadian interests wanted to retain the range for rocket launchings and strategic defence missile testing. To open it to the United Nations would mean either having to duplicate these facilities elsewhere or, worse, asking the United States to relinquish its partnership and leave. Neither scenario presented a very feasible option.[28] Zimmerman and Steacie both agreed, however, that the creation of a Canadian operating facility that attracted international participation was a possibility. Still, both asserted that any such enterprise must remain under Canadian, not United Nations, control.

The meeting's reservations regarding international cooperation provided domestic validation of what was already being perceived at the international level. In late June, the Canadian delegation at the United Nations reported to Ottawa that continued Soviet opposition to space cooperation had rendered the Ad Hoc Committee on Space incapable of accomplishing anything. After the submission of the NRC-DRB technical feasibility study on 27 August 1959, the notion of a Canadian initiative for international space cooperation was dropped. The report, only seven pages long, is completely void of any ambitious statements, perhaps in an effort to dissuade Cabinet from embracing a multilateral space agenda. More importantly, it effectively closed the door for the present on international space cooperation while reinforcing the value of Canada's bilateral relationship with the United States.

Bilateral over International

Attempts at championing a multilateral approach to outer space, albeit with the intent of placing Canada in a favourable position vis-à-vis the American and Soviet domination of space, had failed. The question still remained: if an internationally focused rocket and space strategy were not viable, how should Canada proceed? It had forged a strong cooperative spirit in rocketry and space flight with the United States. Therefore, while it pursued middle-power politics elsewhere during the 1960s, in the exploitation of space, it chose to remain firmly allied to those from whom it could most benefit.

The adoption of this strategy in lieu of internationalism made sense. Senior decision makers within the DRB and the NRC did not expect that Canada would undertake significant space exploration programs on its own, given the low level of attention that such activities had received from government thus far. Canada would certainly participate in the exploration of space, but already it lacked the critical resources to invest in large-scale undertakings similar to those of other countries. Complex space research, orbital research, space stations, and lunar and planetary probes were all perceived as beyond the scope of Canadian financing or necessity.[29] Instead, Canada saw itself focusing on specific projects that delivered unequivocal benefits and encouraged cooperation.

Although no new Canadian space policy was put in place, neither was any policy ratified that might compromise the existing missile, rocketry, and space cooperation efforts with Canada's American partners. Instead, the DRB and the NRC signed a number of bilateral agreements that traded Canadian niche capabilities for general access to broader American systems and technologies.[30] These agreements were critical as both of Canada's main efforts, the DRB space science program and the emerging RCAF Space Defence Program, relied heavily on American support and assured access to outer space.

Several accords were initiated between the DRB and NASA concerning the Alouette-ISIS satellite project between 1959 and 1964. Three of these were letters of agreement directed specifically at the Alouette program, formalizing Canada's first official cooperation effort in the launching of a satellite and exchanging data obtained from it. As well, a memorandum of understanding was put into effect in May 1963 between the DRB and NASA concerning the ISIS satellite project, followed up in May 1964 with a further exchange of notes regarding the results of this program.

Canada also signed a number of agreements related to satellite tracking. In addition to those already in place for both NORAD as well as a program known simply as SPACETRACK, an official agreement was completed in 1960 concerning the location of an American satellite-tracking station at St. John's, Newfoundland, with subsequent amendments in 1962 to update the equipment and convert the existing facility into a minitrack station. This newer design was essentially a modernized version of the original facility but with many improved capabilities. From this accord, Canada gained access to the scientific data obtained from the tracking station and probably also received certain space-derived intelligence from the American collection point. Similar stations were located elsewhere across Canada as increasing numbers of human-made objects entered orbit.

General agreements and exchanges of notes with respect to communications satellites were made between the two countries in 1963 and 1964, the latter aiming to secure arrangements for Canada to cooperate in the eventual creation of a global commercial communications satellite system (GCCSS).[31] At the time, no country had yet achieved a domestic communications satellite capability, and the United States had only just passed its own Satellite Communications Act in Congress in 1962. Canadian participation in this effort led to further involvement with the recently created Communications Satellite Corporation in the United States and later with the International Telecommunications Union and the Interim Communications Satellite Committee. The latter organization was responsible for the establishment of the space segment of the GCCSS, known more commonly as INTELSAT.[32]

An exchange of notes on future Canadian-American cooperation at the Churchill Research Range was completed in June 1965, renewing the recently expired agreement of June 1960 that saw the return of the range to Canadian authority and control.[33] Given the infrastructure already in place and the expenditure required to keep the research range running, it made sense to divide that cost with the Americans rather than have them withdraw and remove most if not all of their onsite resources. Again, the agreement was a demonstration of Canada's desire to maximize its political leverage in lieu of spending to maintain a capability that otherwise might be lost. Such was the prevailing agenda within Canada's bi-nationally oriented space strategy.

Lastly, it is equally important to note that Canada's cooperative space endeavours during this period were virtually exclusive. Between 1957 and 1967, with the exception of some aspects of its satellite communications evolution, Canada signed rocketry or space development agreements solely with the United States and had even retreated from entering into negotiations with France and other European countries on space cooperation. Though interested in exploring various cooperative options with Europe, Canada simply would not enter into any official agreements that might compromise its bilateral arrangements with the United States.

For example, a 1967 proposal for both rocket launcher systems and satellite cooperation with France and West Germany was declined, as it appeared to compete for services already provided to Canada by the United States. Subsequent agreements proposed by these two countries were also stalled, as Canadian officials suggested that signing space cooperation accords with West Germany would be inappropriate, given that none yet existed with closer allies such as Britain.[34] The fact that the DRB, essentially Canada's national space agency, was itself an arm of the Department of National Defence, also generated some concern. Although this presented no problem in dealings with the United States, with whom Canada already had a close defence and security relationship, the DRB's involvement in defence matters probably sent the wrong political signal to European nations. As will be discussed below, this perceived conflict of interest eventually became politically detrimental for both government and defence interests in Canada.

Although Britain had received some special consideration from the Canadian government in space cooperation, this extended only to sharing information and scientific intelligence, as was the case with the telemetry received from the Alouette satellite project, rather than to actual projects or programs. This limitation was partially due to the fact that Britain's rocket and space program was no further advanced than Canada's, and in some scientific and technological fields, the British lagged behind. In the end, Canadian-American space cooperation surpassed all other efforts, including those of Canada's other close allies.

Interestingly, no clear evidence exists to suggest that Ottawa ever took action to purposely distance itself from Washington so as to appease its European partners. Contemporary accounts of Canada's space program tend to make much of the relationships with the European Space Agency and other overseas partners as a form of balance to American influence, yet no such agenda appears to have existed. Dealing with European agencies was simply not in Canada's interest at the time, for they could offer little politically or materially that would assist Ottawa's agenda.

This held true even within international forums. After the failed attempt at instituting international control of space via the UN, Canada maintained only two UN space-related memberships. One was with the Committee on Space Research, whose Canadian membership was held by the NRC's Associate Committee on Space Research throughout most of the 1960s. The second was the UN General Assembly's Committee on the Peaceful Uses of Outer Space, whose Canadian membership was organized through the Department of External Affairs. However, both committees were more politically aesthetic than functional, as space remained a Cold War battleground and was not yet ready for international laws or control. Essentially impotent, neither committee exercised any real influence on Canada's international space agenda until the end of the decade.[35]

International treaties were also held in the same regard. Canada's ratification of the 1967 Outer Space Treaty is often identified as the origin of its non-weaponization-of-space philosophy, but the fact that Canada joined the UN space committees and endorsed various resolutions is often given much more weight than it deserves. The nature and scope of the Canadian rocket and space program at this time, including its significant involvement in defence-related activities, meant that both the 1967 Outer Space Treaty and the 1968 Rescue Agreement made little difference to Canada's overall interests or agenda. They were perceived as having no greater magnitude than other diplomatic ventures. Only after the Canadian space programs had completely transferred to the civilian sector did these agreements receive more attention from the government's oversight committees.

Government Reorganization and Space Politics

In 1960, halfway through Prime Minister Diefenbaker's term, the Royal Commission on Government Organization was appointed to examine the whole of government services, to eliminate duplication of effort and uneconomic operations and practices, and to recommend improvements in decentralization and management. Chaired by John G. Glassco, the Glassco Commission, as it soon became known, offered a number of recommendations concerning Canadian science and technology policy, management, and development. However, events such as the Cuban Missile Crisis and the subsequent federal election of 1963 overtook the commission's work, and its initial recommendations on Canada's space program were not implemented until some years later.

The absence of a clear senior-level political agenda or direction meant that Canada's space advocates were largely left to their own devices in promulgating national interests, research, and development during most of the 1960s. Bilateral agreements continued to be signed on a project-by-project basis, with no real attempt to weave each agreement into a larger Canadian vision, mandate, or agenda. One might argue that this piecemeal approach *was* the agenda, for Cabinet appeared content to let lower-level organizations such as the DRB and the RCAF lead the way as they saw fit or was perhaps too preoccupied with other matters to devote more energy to defining a Canadian role in space. Even direct advice could not greatly change the evolution of Canada's space program. Evidently, government looked upon science and technology in much the same way as it did defence – ignoring them except when they became necessary, and even then only when they were needed to satisfy a political purpose rather than a military, scientific, or technological goal.

Though the DRB, the NRC, and the RCAF repeatedly encouraged Cabinet to approve an overarching national space policy and create a Canadian space agency, the government consistently chose – when it bothered to respond – to retain the status quo. In July 1961, D.C. Rose submitted a proposal to create a national space research organization, but this bore no fruit. Further efforts to solidify official policy after the first launch of the Alouette in September 1962, a time when enthusiasm for space was high within government, also failed to gain increased Cabinet attention. On 3 December, the DRB issued a persuasive position paper arguing for the creation of a Canadian space agency and an official policy to direct future efforts. The study cited the many space activities that were already under way in Canada and the obvious advantage of coordinating government and industry ventures as reasons for creating the agency, but the government still stalled in taking any decision on the issue.

Very likely, the motivation behind the DRB request was twofold. First, the DRB wished to see the birth of a Canadian space agency. Though it enjoyed its position as the primary point of contact for Canada's scientific, engineering, and civilian-oriented space projects and cooperation, it was not a civilian establishment. This became an especially noticeable problem as it expanded its cooperative ventures with countries other than the United States. "Due to DRB involvement in defence matters," one memorandum stated, "its appearance to other nations when participating in such political arrangements ... will have a different connotation than would the participation of a non-military research organization."[36] The DRB was acting as if it were an official Canadian space agency, but it was not and probably should not have assumed this stance. Such activities distracted the DRB from its primary mandate of providing for defence research, a lack of focus that contributed to its eventual demise many years later.

Second, the impending formalization of an official military space program under the RCAF, known as the Space Defence Program, and the competition for scarce assets such as human resources and computing power that would ensue, certainly pushed the DRB to stake an early claim on the future space agenda. Again, playing the role of civilian space agency encouraged it to see itself in competition with an armed service that it should have been supporting. Here, one can also see how the link between national defence and national welfare was obscured and the relationship between science, technology, defence, and government confused. Why did the DRB want to compete against a defence space agenda when it was itself a defence research organization? Obviously, somewhere within the defence science community, the Faustian bargain of accepting defence direction and priorities in order to secure dedicated research funding began to crumble, and the DRB started seeing itself as no longer purely defence-oriented. Though speculative, this would explain why the DRB perceived the RCAF agenda as anathema to its own civilian-oriented priorities. The advent of government reorganization simply complicated the situation for the DRB and the department it served.

Studies produced by the ongoing Royal Commission on Government Organization also identified the serious fragmentation of Canada's space program. Though the commission was seemingly less interested in Canada's military space plans, which were addressed under its reports dealing with the Department of National Defence, it strongly recommended, on paper at least, that Canada's civilian and commercial space interests should not reside under a single authority.

In January 1963, the Glassco Commission focused particular attention on the inefficiency and duplication of effort within Canada's non-military space research efforts and sought recommendations for consolidating them into a single agency. The guidelines for a comprehensive analysis were drawn up, and a subcommittee consisting of specialists from the civilian space research community was assembled to investigate and assess four main areas. As stated in the final report later submitted by the Glassco Commission, these were as follows: all upper atmosphere research conducted by the DRB that was not of direct significance to defence, whether or not it involved the use of rockets; all upper atmosphere and satellite research in NRC divisions, such as that on cosmic rays; the Radio and Electrical Engineering Division's research on meteors and satellites and the operation of minitrack instrumentation; and ongoing research on satellite communications on behalf of the Department of Transport, which would not be encouraged to set up its own facilities.[37]

Interestingly, despite the fact that research had not yet begun on the matter, the commission also recommended that "all non-military space and telecommunications research could be transferred to the Radio and Electrical Engineering Division of the National Research Council."[38] Making recommendations in advance of studies and analysis was common in Canada's review of national science and technology policy throughout this period. The practice corroborates accusations that reviews often arrived at predetermined solutions that were intended to benefit certain parties and agendas, not necessarily national research and development.[39] Still, not even the Glassco recommendations could spark serious interest in better situating the national space policy. In fact, the subject atrophied for another three and a half years.

During this time, the DRB and the RCAF clashed constantly over which of them should take priority in determining the direction of Canadian space activities. The DRB was clearly making headway on its own agenda, with the Alouette I satellite continuing to operate years after its expected life of several days to several months, and with preparations for launch of the Alouette II well under way. In contrast, the planned RCAF Space Defence Program remained largely in its conceptual stage while its proponents attempted to convince Cabinet of its long-term political and military value, all the while dodging challenges from the DRB regarding progress. For every argument advanced by the RCAF, the DRB would counterstrike with an opposing view. For example, the DRB criticized the program's high degree of dependency on and utilization of American technology, and questioned whether it would be compatible with the emerging objectives of Canadian foreign and domestic policy. The reservations regarding compatibility were engendered in part by the realization that a proposed RCAF anti-satellite system would have few non-military applications and that the legality of a peacetime inspection in outer space that might affect satellite operation or endanger human occupants of a manned spacecraft was uncertain and yet to be defined. In response, the RCAF asserted that compatibility was not a problem, for the program might enable Canada to enforce UN resolutions and other international agreements on permissible military activities in space. However, the Pearson government had no wish to take on such a role. Ultimately, the DRB succeeded in making the RCAF program seem politically and technically unviable, and political support for it soon waned.[40]

By the fall of 1964, the RCAF-DRB conflict had reached a new low. Continuing their criticism of the RCAF space plan, a number of DRB staff members regularly advocated in various internal forums that Canada should engage only in purely scientific research programs. As for military space activities, the group argued that Canada should be restricted to the monitoring of US and other military programs by a headquarters group so as to advise the minister of national defence on the implications of any proposals that might benefit from the involvement of the Canadian armed services or the commitment of Canadian financial resources.[41]

Not satisfied solely with internal intriguing, the DRB then went a step further and degraded the value of the RCAF program in a public medium, suggesting that Canada's air force was simply toadying to US coercion by pushing the government for its own military space program. An internal DND appreciation of the growing rift noted that, on 22 September 1964, a Canadian Press release in the opening pages of the *Montreal Gazette* had attributed the following quote to a member of the DRB headquarters: "The possibility that the North American Air Defence Command agreement will not be renewed unless Canada is willing to contribute to costly space defence is being openly expressed by officials here." Printed just three weeks before a critical meeting on military space activities with the United States and Britain, this article was considered significant at Canadian Forces Headquarters (CFHQ), and it very nearly jeopardized an important portion of Canada's national defence cooperation agenda. In a memorandum to Air Vice-Marshal Victor S.J. Millard at CFHQ, Wing Commander L. Birchall, acting director of DAED (Directorate of Advanced Engineering and Development), stated, "these military space meetings are normally difficult for the Canadian delegation because of the paucity of Canadian projects in this field. In view of the opinions expressed in this Canadian Press release, the forthcoming meeting is likely to become even more difficult. It could, in fact, lead to a major reservation on the part of the US as to the real RCAF interest in space matters."

Worse, perhaps, was the fact that the Canadian delegation to such meetings normally consisted of personnel from both the RCAF and the DRB. "It is therefore of great importance to the effectiveness of future DND participation in space projects," Birchall stressed, "for the Canadian delegation to be able to demonstrate solidarity on the Canadian military position with respect to future space research and development."[42] A forced truce did much to control the damage, and when the members of the Canadian delegation, which consisted of Group Captain Peter (CFHQ rep) and R.C. Langille (DRB rep and Canadian national leader), sat down at the table with their American and British counterparts, they spoke in harmony and were prepared to discuss the advances in Canada's military space program. Although the meeting passed without incident, the damage to DRB-RCAF relations in space programs was more lasting. It is very likely that this acrimony simply compounded already significant mutual

FIGURE 22 R.C. "Bob" Langille led the development of Canada's Cold War radar research, served as DRTE director general from 1969 to 1971, and was later the advisor on space policy to the assistant deputy minister, Space Programs. Communications Research Centre, 69-18206.

distrust between the two organizations, encouraging them and their competing agendas to go their separate ways.

The National Space Study

The ongoing debate over the future of Canada's space program finally reached a watershed. By 1965 Canada was at an impasse and faced the possibility of being unable to realize any further program achievements beyond those currently under way. Though Canadian rocketry and space activities had demonstrated some degree of success thus far, the country's piecemeal approach to its program and the splintered management of its projects caused serious political strife, wasted valuable technical and human resources, withered funding, and hampered longer-term national science and technological objectives.

No longer could the various proponents of space exploration and exploitation depend on the traditional relationship in which individual scientists or organizations dealt directly with those in government who controlled the funding. That era had passed. The advent of space exploration, and the race between the superpowers to put a man on the moon by the end of the decade, clearly demonstrated the enormous impact that science and technology could have on wider political issues, government programs, and social values. It also brought public attention to the ever-increasing cost of "big" science and invited searching questions at home about how Canadian science and technology were governed, managed, financed, and directed.

During its final days of existence, the short-lived Science Secretariat turned its attention to Canada's rocket and space program and policy. In May 1966, just as the Science Council of Canada Act was receiving assent in the House of Commons, the secretariat commissioned the National Space Study, a comprehensive analysis based on the parameters identified in the Glassco Report. One would have expected the RCAF and the DRB to receive equal attention in a national-level assessment such as that coming from the Glassco Report, but as the report itself explained without any elaboration, it sought to address non-military issues only. Though evaluating a national space policy primarily in civilian terms made perfect sense – after all, Canada was governed by a civilian democratic authority – this decision was a determining factor in ensuring that the DRB's science-oriented agenda took precedence over the more technologically focused RCAF Space Defence Program.

A group of three distinguished scientists – Peter A. Forsyth from the University of Western Ontario, P.A. Lapp from de Havilland Aircraft of Canada, and G.N. Patterson from the University of Toronto – were assigned to the Science Secretariat National Space Study group under the leadership of John H. Chapman, one of Canada's leading space engineers.[43] Already a respected and well-known figure in Canada's fledgling space technology community, Chapman was a skilled radio scientist and liaison officer then working in the Defence Research Telecommunications Establishment. He had overseen most of Canada's bilateral space cooperation with the United States since 1958 and was largely responsible for the overall success of the Alouette and ISIS satellite projects.

The study group, also referred to occasionally as the Ad Hoc Task Force or the Chapman Task Force, collected data by requesting it from various stakeholders in Canadian rocketry and space activities and through numerous hearings

FIGURE 23 Peter A. Forsyth joined the Radio Physics Lab in 1952 and later played a central role in the early development of Canada's space programs and policies alongside John Chapman. Communications Research Centre, 53-RPL-0321.

and interviews. Submissions arrived from universities, industry, government departments and agencies, technical associations, and professionals associated with Canadian space projects. In addition, between 30 June and 31 October 1966, the study group presided over hearings at Halifax, Quebec, Montreal, Ottawa, Toronto, London, Winnipeg, Saskatoon, Calgary, Edmonton, and Vancouver. In total, 112 briefings and presentations were made during these hearings.[44]

These exchanges generated thousands of pages of ideas, options, technical notes, and status reports. In terms of looking ahead to the future, three points repeatedly surfaced throughout the group's investigation. The first was the overwhelming consensus that there was an immediate need for a centralized organization to oversee all space activities in Canada. Second, the current success in telecommunications research should lead toward the development and deployment of Canadian satellites for domestic telecommunications by 1970 or 1971 at the latest. And third, if Canada wished to ensure its access to outer space, an indigenous satellite-launching capability was required.

The first and third points were old arguments. The Canadian space community had advocated for the creation of a Canadian Space Agency since nearly the beginning of Canada's involvement in the space age, and the indigenous launch capability issue had first appeared on the agenda of the Associate Committee on Space Research at its first meeting in 1959. However, the second point was new and almost certainly suited the agendas of both those who were conducting the review and those who were being interviewed. Both John Chapman and Peter Forsyth had a long relationship with the government experimental communications satellite program and were keen to continue political support for their own research and development in this field.

In fairness, however, it was a logical priority for the committee to include and was a feasible goal around which to base an official policy, program, and agency. Canada had achieved considerable success in developing satellite communications technology throughout the 1960s; compared to the rest of its fledgling program, this area was also the most advanced. There was little doubt that it would receive considerable focus in the study group's final report.

Preliminary copies of the Chapman Report were distributed to members of the Science Council of Canada in advance of an official meeting planned for 16 January 1967 to discuss the group's findings. A thick and detailed document at 258 pages, it was divided into three sections. The first described government, university, and industrial space projects created and planned for between 1961 and 1971. Every aspect of Canada's space program was included in this portion, from detailed descriptions of scientific experiments and equipment to budgetary summaries and human resource allocations. The second part of the study included the texts of space agreements, memorandums of understanding, and other international arrangements affecting or concerning Canada and in effect as of 31 October 1966. The last part, an appendix to the first section, presented a detailed and comprehensive review of Gerald Bull's High Altitude Research Program then under way at McGill University.

The Chapman Report was well received by the Science Council of Canada at its official tabling in January 1967 and was subsequently published again several times, first in February as *Upper Atmosphere and Space Programs in Canada: Special Study No. 1* of the Science Secretariat. It also became a topic of debate and decision at follow-on Science Council of Canada meetings in March, May, and June.[45] This was followed in July 1967 by another officially published version, Report No. 1 of the Science Council of Canada, which was titled *A Space Program for Canada*. It was on this last document, a concise version of the Chapman Report, that Cabinet largely based its plan for future space activities in Canada.

Conclusion

The difficult and disjointed nature of Canada's space policy development during the first decade of the space age was reflective of wider events taking place between science and government during the 1950s and 1960s. At the highest level, Canada had no clearly defined national goals upon which new science policy could be based other than those related to economic welfare as laid out by the Economic Council of Canada. Speaking before the Senate Special Committee on Science Policy in 1966, Omond Solandt, then chairman of the Science Council of Canada, observed that, although many *statements* appeared in various sources discussing national goals, almost none of these were codified. Furthermore, they were not very useful politically as frames of reference for real policy development. "Many of them are mutually conflicting," Solandt stated. "You cannot have them all at once."[46] This was particularly true in the rocket and space program, where success depended on mutually supportive, not exclusive, aims and objectives.

Unable or unwilling to generate new distinctive policy for science and technology during this period, both Conservative and Liberal Cabinets defaulted to the traditional approach of supporting mission-oriented scientific and technological research or applying DND's science and technology applications to other political, defence, or social objectives.[47] This is certainly how space activities were treated prior to 1967. One could argue, for example, that the Alouette-ISIS satellite project existed, not because Canada had a long-term goal to develop domestic satellite communications, but because Ottawa chose to support a single scientific application in order to gain significant political, technological, and defence benefits from its American ally. True to form, rather than expend resources and effort on an elaborate and all-encompassing space program, the country sought to consolidate and concentrate where it could, trading scientific and technological innovation for direct applications that provided economically tangible benefits. Until 1967 at least, a long-term vision for space was abandoned in favour of short-term returns.

It is little wonder, then, that at the end of 1967 Canada still had no officially ratified space policy, and the government did not seem prepared to embrace the preparation of one. Probably, it simply did not perceive a need for it, though its suborganizations desperately required some top-level direction. Even when strategic guidance for space activities finally did arrive with the publication of *A Space Program for Canada,* the Science Council admitted that the program's identification as a priority was "partly determined by circumstances, and partly by present importance," not because it was connected to the achievement of national goals.[48]

Such explanations for the absence of a proactive government agenda may seem frustrating, but there is simply no evidence to suggest that Ottawa was either intentionally stalling or devoting serious thought to the subject. During this period, its national priorities simply lay elsewhere, and Cabinets were content to allow the conquest of space to proceed at whatever speed its various subordinate organizations could dedicate to the effort.

American historian Robert L. Heilbroner wrote in 1962, "adrift on a furious current of technology, we allow ourselves to be swept along, trusting to the blind forces at work to bring us safely to some unknown but unquestioned destiny."[49] This certainly seemed to be the case with Canada's space program until 1967. In that year, however, Cabinet finally realized that, from the perspective of both economics and national welfare, Canada could no longer afford to "expose herself to the degree of economic and technological dependence" that characterized the national space program.[50] At some level or in some program, Canada needed to move forward and declare technological independence in support of its own interests. During the following year, the first steps were taken to realize a level of technological independence, as changing circumstances became important if not critical to national science goals and a new agenda for Canada in outer space.

4

Forging a Spacefaring Nation: The Alouette-ISIS Program

ALTHOUGH THE AUTUMN OF 1962 was perhaps most historically notable for the Cuban Missile Crisis, in the days prior to the worst part of that episode, Canada experienced a brief respite from the daily issue of bad political news through an achievement in space technology. On 28 September, as a cool wind blew in to the California coast from the otherwise tranquil Pacific Ocean, an American-built Thor-Agena rocket left its pad at Vandenberg Air Force Base carrying the first Canadian satellite, affectionately named Alouette, into orbit.[1] For the rather technologically nascent Canadian space community, this was a remarkable accomplishment. Less than four years earlier, the country's first satellite project had been a merely academic defence science proposal, but as Alouette I lifted into the evening sky, Canada formally joined the elite club of space-exploring nations.[2]

The launch of Alouette marked a unique political and technical apex in Canada's Cold War history. For a brief period, Canada surpassed all other Western allies and nations save the United States with a demonstration of advanced technical capability then matched only by the Americans and the Soviets. Not even the launch of the British experimental satellite Ariel I in April 1962 attracted as much attention as Alouette I; nor did it produce as plentiful a return in data. Furthermore, Alouette I far exceeded its life expectancy of six months by remaining operational for a decade and in turn prompting long-term political and financial support that spawned Canada's only full-scale space project during the 1960s – the Alouette-ISIS program. Arguably, it was among Canada's best technological achievements in the early post-war era.

As remarkable an attainment as it was, the Alouette-ISIS project, and in a greater sense Canada's role in the evolution of space science and technology, has yet to be placed within the larger context of the country's history.[3] Canada emerged from the devastation of the Second World War with a vigorous industrial economy that had remained largely unscathed by conflict. Between 1945 and approximately 1970, following generally accepted principles of innovation theory as defined by Harvard economist Joseph Schumpeter (*Business Cycles*, 1939) and fellow economist Jacob Schmookler (*Invention and Economic Growth*, 1966), the country was poised to secure an early advantage in the fourth innovation wave – identified as the cluster of advances surrounding the development of computers, solid-state electronics, digitization, miniaturization, and synthetic materials.[4]

Some historians have argued, however, that this initial advantage was offset by the absence of an industrial tradition of scientific research and experimental development, especially in those areas needed to drive innovation in modern digitization and miniaturization.[5] Though this view is still generally accepted, more recent scholarship on Canadian science and technology history is beginning to challenge it. For example, studies of early Canadian computing development have shown that considerable national technological competence existed in several fields even though the country lacked a large domestic market, such as that in the United States, through which to diffuse such technologies and encourage their further development.

Nonetheless, if post-war Canada lacked a technologically innovative culture, how did it manage to reach outer space before many other Western nations with similar economies? Was the Alouette-ISIS project simply an anomaly in historian J.J. Brown's risk adverse interpretation of Canada's post-war technological evolution, or was it another example of economist Jacob Schmookler's demand-driven incremental innovation as later supported by historians Christian DeBresson and John Vardalas?[6] As former DRTE scientist LeRoy Nelms recently noted in a brief overview of Canada's space history, "to this day, the whole story of the space program and its impact on DRTE, DRB, and DND is not entirely clear."[7] A more detailed study of Canada's Cold War Alouette-ISIS program is therefore germane to defining the relationship between science, technology, defence, government, and society during this period of inventiveness.

Finally, a deeper analysis of Canada's entry into the space age is also necessary if the country's early efforts are to be assessed within the context of other general histories of space exploration and discovery. For example, though Canadian advancements in this field were widely recognized and to some degree even depended upon by the United States, subsequent American histories do not strongly identify this relationship. This is not an intentional slight but more probably arises from the lack of easily accessible Canadian space histories from which to draw a useful historical reference.[8]

The DRB Topside Sounder Project

The idea for Canada's first satellite evolved within the space science community of the Defence Research Board and select Canadian universities during the early 1950s, but it came to fruition after the United States National Academy of Sciences published an international call for scientific experiments to be flown on a satellite. The academy was tasked by the US government to coordinate both internal and international efforts in ionospheric research; thus, on its behalf,

Lloyd Berkner issued an open invitation in early 1958 to the Western scientific community to submit proposals for satellite-borne experiments that could explore the "topside" of the ionosphere. A number of countries returned submissions, one of which was a rather clever idea from the DRTE in Ottawa.

Within the DRTE, Eldon Warren, Colin Hines, John Chapman, and a few others had contemplated the idea of a Canadian scientific satellite for some time, as had Peter Forsyth and some of his colleagues working at the University of Saskatchewan. However, when the DRB began officially canvassing its personnel for potential interest, satellite experiments were not an easy sell. Hines of DND's Radio Physics Laboratory (RPL) later recalled that the DRTE chief superintendent James C.W. Scott "called me in and asked if I wanted RPL to do satellite-borne studies. I replied that in due course I did ... He told me there was a proposal for DRTE to build a baseball-sized chunk of equipment for topside sounding, to be carried aboard someone else's satellite ... I simply replied that, perhaps, it would be a good thing for us to do in another year or so."[9] Hines politely declined the offer in order to properly address other commitments within RPL, but Scott turned the project over to the Electronics Laboratory for consideration.

Eldon Warren of the Electronics Laboratory, then currently involved in "bottom side" sounding of the ionosphere, immediately warmed to the idea. He developed the concept along with John Chapman and other talented scientists and engineers in the laboratory and organized a proposal for DRB senior review. The design, considered both original and very technically sound, was subsequently approved for submission to the National Academy of Sciences.

In their proposal, the DRTE engineers reasoned that their submission would meet with greater approval if it demonstrated advanced engineering capabilities. Thus, instead of submitting a single-frequency investigation experiment as other organizations did, the Canadians proposed to build a second-generation satellite that could employ a sounder capable of sweeping through a range of frequencies. In September 1958, when the interested agencies met at the American Space Science Board's Working Group on Satellite Ionosphere Measurements in Boulder, Colorado, they agreed that the DRTE concept indeed featured advanced capabilities that none of the other parties had considered.[10]

A special independent meeting to consider the topside sounder experiment in detail was called by H.G. Booker of Cornell University in October 1958, attracting the attention of at least seven interested groups, including the team from the DRTE.[11] Again, their proposal met with a favourable response and was eventually selected as the preferred option. Unfortunately, just a short time afterward, the academy cancelled its participation in the venture, and the offer to the DRTE "farm team" was rescinded.[12]

Figure 24 James C.W. Scott was an RCAF squadron leader during the Second World War who later became a leading world expert in ionospheric studies. He was acting chief superintendent of the DRTE between 1955 and 1959 while Frank T. Davies was seconded to Ottawa and served on the Canadian Joint Staff in London, UK, from 1959 to 1963. Communications Research Centre, 65-9873.

Undeterred by this setback, the DRTE approached interested parties in the United States in late 1958, this time targeting the US Department of Defense. Former DRTE scientist and Communications Research Center historian LeRoy Nelms has written a popularized account of their visit to the Pentagon as follows:

> It is said that they [Scott and Warren] were greeted at DoD by a big Texan USAF Colonel who listened attentively, Cowboy boots on his desk, and smoking a large

> cigar, as these boys from the far north explained their proposal. When they had finished, he put his feet down, snuffed out his cigar and said, "Sure we'll launch for you. But there's a new agency just starting up – called NASA – who are supposed to do international space research projects. Probably won't amount to anything but you'd better go see them first. But if they aren't interested y'all come right back and we'll look after you."[13]

Finally, Scott and Warren approached the newly created US civilian space authority NASA in early 1959. Again, the DRTE engineers impressed their potential sponsor – the fact that they already had a carefully planned proposal in hand from their previous attempts certainly helped – but apparently NASA received it with a degree of skepticism. The NASA scientists were concerned about the craft's power supply and antennas. In the Canadian design, power would need to be continually generated, and four robust antennas measuring between twenty-two and forty-six metres in length yet weighing no more than 4.5 kilos would be deployed. Some at NASA felt this was unachievable and argued that the design be rejected. Others believed that this criticism was no more than a stalling tactic to ensure that a similar American-sponsored project then in development, known simply as S-48, would be launched first. Whatever the case, the Canadian team got its chance in the end, and though NASA later admitted that it initially expected the satellite to operate for no more than a few hours, it was glad to be proven wrong.[14]

Still, the proposal fit well into NASA's emerging mandate, and after some negotiation the two organizations agreed to cooperate on the Canadian topside sounder satellite experiment. Both countries made a joint announcement of the arrangement on 20 April 1959, with an official exchange of letters between the DRB and NASA following on 25 August, 18 November, and finally 6 December 1959.[15]

With the advent of orbital capability, Canadian defence scientists once again had the opportunity to expand their investigation of the upper atmosphere. Through the DRB, DND supported these endeavours, which were seen as a positive way of acquiring advanced space technology then considered critical to defence. From the mid-1950s onward, knowledge of the earth's ionosphere played an ever-increasing role in the design of modern command and control architectures, now largely based on wireless radio communications. As well, the development of advanced weapons systems and defences was increasingly dependent on technologies similar to those created for space systems, such as solid-state electronics, miniaturization, and even computers. Given the military value of the material derived from satellites – namely, imagery and scientific data – the defence scientists had little difficulty in convincing their leaders of the need for continued research and financial support. The possibility that these innovations might later be diffused from the military into the civilian economy was another factor that generated sustained support. The government saw space technology as an industrial opportunity. The defence scientists were only too happy to present it in that way.

Project S-27: Initial Planning

While making formal political arrangements with Canada, NASA asked the Central Radio Propagation Laboratory (CRPL) of the National Bureau of Standards to examine the proposals for their scientific merit and engineering feasibility. In considering the Canadian proposal, the CRPL suggested that it might be rather ambitious for a first attempt at examining the ionosphere. It suggested that a fixed-frequency system should be launched as a first-generation experiment while the DRTE swept-frequency system proposal was developed as a second-generation satellite. All parties agreed to this and planned for the Canadian launch some time in 1963.[16] For the time being, NASA simply designated the DRTE satellite experiment as Project S-27. Renaming the satellite as Alouette I came later.

In January 1960, NASA received a joint proposal from an American organization named the Airborne Instruments Laboratory and the CRPL for a fixed-frequency satellite. The former was to design, construct, and test the satellite payload, and the latter was to provide scientific supervision and analyze the resulting data. Work began on this new project, designated S-48, on 9 May 1960. It would supposedly precede the DRTE satellite, but because it was similar in objectives and technique, it would in fact simply complement the Canadian effort. The main difference between the two satellites was in the instrumentation. The American-designed S-48 emphasized the study of meridian cross-sections through the ionosphere, and it employed Canadian and American telemetry stations along the 75°W meridian. It also had a low resolution and a fast profile acquisition rate, employing six fixed frequencies providing a downward pulse transmission and echo reception in the three- to nine-megacycle range.

By contrast, the Canadian S-27 satellite emphasized the investigation of polar, Arctic, and auroral effects that produced the complex ionospheric conditions existing over Canada. To achieve this, telemetry stations were established at specific points across the country to collect data transmitted from the satellite. As well, the S-27 differed from the S-48 in that it employed a high-resolution and slow-profile acquisition rate.[17]

The scope of the Canadian experiment soon attracted interest from the United Kingdom, which expressed a desire to participate in the topside sounder

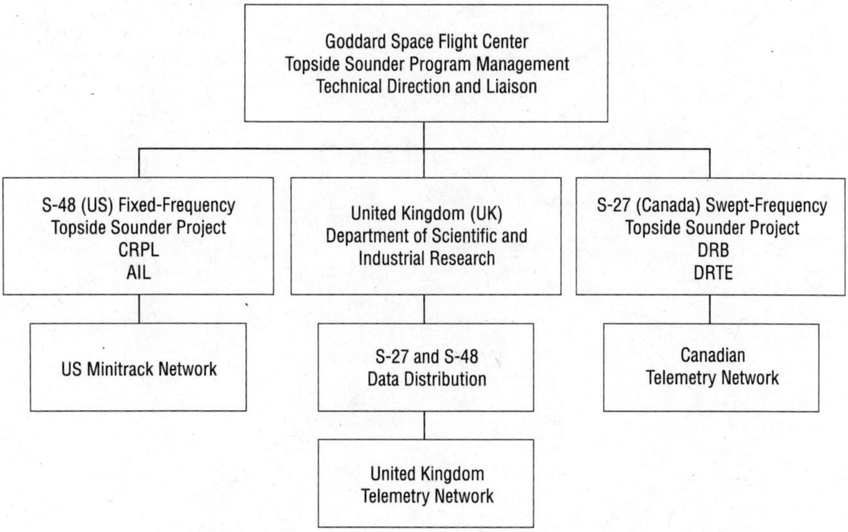

FIGURE 25 Project management and oversight for Canada's first satellite project.

program. In return for access to data, the United Kingdom offered the use of three of its ground telemetry stations – one at the Falkland Islands in the South Atlantic, one at Singapore, and one at Winkfield, England – for the collection and distribution of satellite data. NASA and the DRB accepted this offer, and the United Kingdom officially joined the program later that year.

The topside sounder project came under the overall management of the Goddard Space Flight Center, which coordinated the efforts of the countries and organizations involved (Figure 25).[18]

Design, Construction, and Testing

In the late 1950s, space systems design, construction, and testing were unproven processes. Little was known about space as an environment or about how various materials might survive in its vacuum over long periods of time. Likewise, modern management of research and development was still in its infancy, and as scientists and engineers were working out a viable process for coordinating large-scale technology development, their leaders were looking for organizations that could learn, adapt, and sustain adaptation to achieve long-term goals.[19] It was within this context that the DRTE was expected to produce Canada's first full-scale satellite.

Understandably, the approach to designing it was conservative. The original concept called for a spacecraft that resembled the first American satellites, being no larger than a grapefruit or a basketball. As Project S-27 evolved, however, so

FIGURE 26 Alouette I undergoing tests in the thermal vacuum chamber at the Canadian Armaments Research and Development Establishment at Valcartier, c. 1961. Communications Research Centre, 61-RPL-0142.

too did the size and complexity of the satellite. Though the DRTE Electronic Lab team originally wanted it to play only a single role – to measure the state of the ionosphere directly below it as it orbited the earth – the designers added three more experiments as the project progressed. These included a sounder receiver to measure cosmic noise, a frequency receiver for "listening" to radio noise in the range of one to ten kilohertz, and an instrument to measure primary cosmic ray particles such as electrons, protons, and alpha particles outside of the denser portions of the earth's atmosphere.[20] In the end, Alouette resembled an oblate spheroid of slightly more than a metre in diameter, with a height of approximately 0.86 metres. Its total weight was nearly 145 kilos.[21]

Like most satellites of the period, Alouette featured a clamshell-like design. It had four main components: the structure, the spacecraft electronics, the antennas, and the four-experiment payload. Its internal backbone consisted of a pair of thrust tubes, one above and one below, with a pair of circular structure discs between these, which served as the mounting areas for the electronic components of the four experiments. The four erectable antenna units

FIGURE 27 An Alouette engineering model is examined by C.A. Franklin, manager of electrical systems, R.K. Brown, manager of the spacecraft team, and engineer J. Barry, c. 1961. Communications Research Centre, 61-RPL-0479.

(described in more detail below) were also housed between the two discs. Surrounding the centre of the structure was the solar cell shell that consisted of a pair of spinnings upon which were mounted support channels to carry the flat solar cell panels. Inside the spinnings were diaphragm rings that served the twofold purpose of stiffening the spinning and providing the attachment between it and the structure discs.[22]

The structure of the satellite, which held all the other components together, posed the largest design challenge. It would have to withstand the violent vibrations of launch and the vacuum and radiation of space yet still collect and return its data. Several building materials were considered, including sophisticated options such as micarta, polyurethane epoxy, Teflon, aluminized Mylar, and unbonded glass fibre paper, and even some not-so-sophisticated materials such as commercially available brown wrapping paper. To cope with the vacuum conditions, the DRTE engineers had to avoid materials with high partial pressures that sublimed easily. As a result, aluminum was chosen as the primary material for the body of the satellite, held together with steel and stainless steel fasteners.[23] The satellite structure was completed by a cap at either end to prevent sunlight from striking through to the interior and overheating the electronic components there.[24]

Temperature, or more precisely temperature control, affected every aspect of the design. Everything, from shape to material used to launch times, was considered to ensure that the satellite and its precious payload experiments remained within acceptable tolerances during launch and orbit. Many spacecraft had swiftly expired in the harsh environment of space as one side literally cooked while the other side froze. The DRTE engineers were most concerned about their satellite surviving long enough to return useful data to scientists on the ground.[25]

It was expected that the period during and shortly after launch would be most critical for the satellite's life. The intended launch vehicle was an American two-stage Thor-Agena booster. Once the Agena upper stage had separated from the Thor lower stage and initiated its own burn, it would shed its payload shroud, leaving Alouette firmly attached yet totally exposed to the vacuum of space for two and a half minutes. During this time, Alouette would not be spinning and therefore solar heat could not be equalized around its shell. Thus, in an effort to diminish its exposure to heat, it would launch and ascend in the earth's shadow, but even then the problem of aerodynamic heating remained as Alouette headed toward outer space.[26]

The shell of the structure was designed so that the power plant was partly located on the outer surface. Alouette's outer skin was almost completely covered by 6,480 small solar cells arranged in groups of 45, which charged the batteries located within it. To provide adequate charging currents regardless of the satellite's orientation with respect to the sun, the same number of solar cells had to be consistently exposed at all times.

Attached to the cells with an epoxy-based adhesive were paper-thin chips of glass – a special non-reflective and spectrally selective coating that admitted ultra-violet light but not the infrared light that could heat the spacecraft. In a sense, both cells and coating acted as thermal insulators but still let the much-needed light through. Together with the satellite's spin, the glass covers would keep the cells within a temperature range of –20°C to +50°C. In operation, the spacecraft temperature sometimes rose as high as +75°C.[27]

This design also protected the solar cells from micrometeorites and harmful radiation.[28] Though it also acted as a constraint and influenced the semi-spherical shape of the satellite, the power requirements of the payload dictated its overall surface-area size. Finally, the shape contributed to temperature control. That Alouette would have a slight spin in orbit, roughly two revolutions per minute, not only reduced its oscillation but also helped it avoid the danger of getting "hot spots" from overexposure to direct daylight.[29]

The design of the interior electronic components highlighted some of the technological challenges still faced during the early 1960s in advanced systems. Although extensive miniaturization of the electronic content was not considered necessary for Alouette, it was nonetheless desirable if for no other reason than lessening the weight of the craft and potentially making more room for the onboard experiments. Employing vacuum tube technology meant possibly providing more power but doing so at the expense of reliability, higher power consumption, greater weight, and ultimately the need for more space in an already small satellite. Instead, the Alouette team sought out the most modern electronics available, using then ultra-modern solid-state transistors that provided less power but were more reliable and gave greater semi-conducting efficiency.

The antennas initially posed a particular challenge for the engineers but in the end resulted in a novel and uniquely Canadian solution. The Alouette design called for four erectable sounding antennas: two crossed dipoles 45.72 metres from tip to tip and two crossed dipoles measuring 22.86 metres from end to end. All four antennas were designed to extend in a transverse plane at the centre of the satellite exactly ninety degrees to one another and had to be housed within the satellite in such a manner that the payload would fit into the confined payload bus fairing atop the rocket. They would not fit inside the satellite, and they were too fragile to fold and pack alongside the main structure. Further, their multi-jointed nature meant that any one of them could fail, resulting in a partial deployment or even no deployment at all. All the antennas had to deploy perfectly for the satellite to complete its assigned task.

The solution to the problem derived from a tool conceived and developed by G.J. Klein, an engineer with the National Research Council since the Second World War. A taped length of spring steel previously heat treated and opened flat was wound on a drum and placed within an antenna assembly with a guide sleeve and an electro-static shield. Altogether, the antenna assembly unit was no more than about a foot long, and it fit comfortably into the satellite structure housing. Once in orbit, the antenna deployed by pulling the spring steel out of its storage drum by means of a simple drive belt, and once guided through the antenna sleeve, the metal tape sprang back into its natural tubular shape, with about 180 degrees of overlap. Even at 45.72 metres, the antennas proved extremely robust, with considerable bending strength.[30] The tool, later known as the storable tubular extendable member (STEM) system, worked so well that it was employed on nearly all subsequent Canadian and American satellites and spacecraft throughout the next two decades.

Testing the STEM system on the ground was likewise a difficult proposition. NASA's philosophy was that a prototype identical to the flight version of the unit had to survive a program equalling 150 percent of the highest expected design loads. As well, the satellite had to pass several vibration tests in which its dynamic balancing was proven. Also, with Alouette and subsequent Canadian-built satellites employing STEM technology, the antennas had to be tested, which proved a real challenge in Earth's gravity. An instrument known as a full extension rig was employed to extend the antennas along a series of hung carriers, while engineers measured rate of extension, drive motor current, and voltage.[31]

So much of the mission depended on the successful deployment and operation of the antennas. To further ensure their success, the design was flight tested in June 1961, when a pair of the STEM were mounted in the nose cone of a US-built Javelin rocket and launched to satellite altitude. Though the test indicated that some improvements were needed, it was generally very successful and proved for a final time the practicability of the STEM design.

Launch and Results

Canada's first satellite project remained on schedule, and the completed spacecraft was moved to Vandenberg Air Force Base, California, in the late summer of 1962 to undergo final checkout and transfer to the launch pad. Alouette was scheduled for a night flight; the planned launch window of 11:30 to 1:30 on 28-29 September 1962 was chosen as it allowed the mission scientists to get as many soundings as possible on the first few orbits after liftoff before Alouette was fully exposed to the sun, just in case it suffered a catastrophic malfunction or failure when it heated up for the first time. This flight plan also enabled the scientists to ensure that the onboard passive temperature control system was working by slowly exposing the satellite to progressive amounts of sunlight with each passing orbit.

Alouette was launched just before midnight. John Chapman, head of the DRTE Alouette team, later remarked, "I had my fingers crossed, my legs crossed, and everything else crossed. At that time, there was still a fifty percent chance of failure in launchings."[32] However, Chapman and his team had nothing to worry about. The launch took place without difficulty and lit up the evening sky as the American booster lofted Alouette into orbit.

All aspects of the flight were closely monitored. Prior to launch, NASA had arranged through the US Navy to have a tracking ship positioned in the Indian Ocean to monitor Alouette's antenna extension, but at the critical moment, there were equipment and operator troubles aboard the ship. As a result, the Canadian launch team was kept in suspense for some time until a report arrived

FIGURE 28 The launch of Alouette I on 29 September 1962. Canada's first satellite was carried into orbit atop an American Thor-Agena rocket launched from Vandenberg Air Force Base in California. Communications Research Centre, 63-7102.

FIGURE 29 John H. Chapman at Vandenberg Air Force Base, California. Behind him are the rocket and payload bus carrying Canada's first satellite, Alouette I. Overseeing the national space study in the mid-1960s, Chapman exerted the greatest political influence in Canada's Cold War space program. Communications Research Centre, 62-6586.

from a ground tracking station in Johannesburg verifying that the antenna had fully extended.

Nevertheless, the satellite that some NASA scientists had felt would last but a few hours went on to surpass all expectations. Within weeks of the launch, scientists were flooded with detailed data on ionospheric structure collected by Alouette. Having optimistically planned for a three-month mission (Alouette was designed for a nominal lifespan of one year), the scientists collected and processed as much data as possible, but it soon became apparent that the satellite was functioning well and would continue to deliver data for some time. When Alouette passed its first birthday in space, NASA and the DRTE were impressed. When it celebrated its fifth anniversary in orbit, both groups were simply amazed. Even then, no one guessed that it would function for another five years, finally shutting down only after being decommissioned by ground control a decade after its launch.

The success of Alouette was certainly an achievement of which the Canadian defence and scientific community could be proud. Compliments came from most scientific organizations but particularly from the skeptics at NASA and the National Academy of Sciences who had earlier questioned some aspects of the mission. The admiration for Alouette was stated very succinctly in the 1963 publication series from NASA's Goddard Space Flight Center. As one group of authors wrote, "The success of the NASA program of topside ionosphere studies is evidenced by the considerable amount of knowledge obtained from Explorer VIII, Ariel I, and high altitude rocket soundings. Perhaps the most spectacular of these accomplishments to date has been the Canadian swept-frequency topside sounder, Alouette, which will probably yield more data about the upper ionosphere ... than all the other programs combined."[33]

The experimental satellite project also demonstrated what Canadian defence science could achieve when given a clear mandate and the resources with which to fulfill it. However, the design and construction of Alouette should not be perceived as simplistic or something that was easily repeated. The initial concept in place when Scott was chief superintendent called for a very small payload. But, as Colin Hines noted many years later, the political and financial headaches grew with the project: "I don't recall when the baseball expanded into a basket ball, as it did in mid preparations, or then to the ultimate size of Alouette itself, requiring all the time more and more resources. Frank Davies replaced Scott as Chief Superintendent, DRTE, part way through, and from time to time bitched about this Albatross Scott had hung around his neck. The program ultimately took over EL [Electronics Lab], all of the finances and manpower that could be pulled together inside DRTE, and ultimately required massive subsidy from the DRB itself."[34]

Undoubtedly, Alouette came at the cost of approximately $3 million, and it drained resources, scientists, and engineers from all other projects at the DRTE. Those not involved with Alouette were understandably somewhat resentful of the impact that the high-profile satellite project had on other quieter, less spectacular though fundamental defence science research activities at that time. Some within the DRTE later suggested that the success of the project ultimately contributed to the demise of the DRTE in the late 1960s, when it was permanently transferred out of DND to the Department of Communications (DOC).[35]

Nevertheless, the data returned from Alouette about the ionosphere, which proved to be extremely valuable to future defence communications research, also revealed that there was much more to learn about the upper atmosphere than at first realized. That required additional experiments involving other

FIGURE 30 R. Uffen (right), DRB chairman, and John H. Chapman with the first of the ISIS project spacecraft. Communications Research Centre, 68-16577.

ionospheric parameters, "which could only be satisfied by subsequent satellites."[36] Canada's defence scientists were keen to build and launch those satellites.

The ISIS Program

Following the technical and scientific success of Alouette I, an agreement was reached between the United States and Canada on 10 January 1963 to embark on a joint program of launching an additional four satellites devoted to ionospheric research. Known as the international satellites for ionospheric studies (ISIS), the four planned satellites were designated Alouette II (also known as ISIS-X), ISIS-A, ISIS-B, and ISIS-C. They were to be designed and constructed in Canada and launched by NASA at intervals between 1964 and 1969 during a period known as the half cycle of sunspot activity.[37]

The ISIS project would consume most of the DRTE's resources, but the agreement to go ahead depended on the stipulation that Canadian industry be employed to its fullest extent in the process. This clause was inserted in order to stimulate Canada's limited technical and industrial base during the 1960s, so

FIGURE 31 ISIS-A mounted in its payload bus for mating to the launcher. Communications Research Centre, 69-18707.

that by the end of the decade, a skilled industry for spacecraft design and construction would potentially exist in the country. This goal was difficult to meet through the construction of only four satellites, but proposing such national objectives was necessary to convince Ottawa to commit itself both politically and financially to the task and provide for a special parliamentary vote to secure initial funding for the new program.[38]

Planning and working groups for the project gathered in early 1963. An overall plan was developed, with the working group generally advocating for higher inclination orbits than Alouette I had achieved so as to explore a greater range of altitudes. As well, direct measurement experiments were to be included in the ISIS satellites so that mission scientists could collect data on the immediate environment surrounding the spacecraft. Also desirable were more detailed measurements of the charged particle influx into the ionosphere and of its relation to the auroral processes and to the production of ionization.[39]

Work on ISIS-X/Alouette II began in March 1963. Briefings were held to acquaint industry with the program, after which DND and the Department of Defence Production selected companies for the project. The RCA Victor Company of Montreal was chosen as the prime contractor, and de Havilland Aircraft of Canada, in Toronto, was selected as the associate prime contractor. To prepare these companies for satellite manufacturing, their first assignment would be focused on training personnel through the construction of Alouette II. Industry personnel were requested to report to the DRTE at Shirley's Bay near Ottawa for indoctrination and to set up the shops where the satellites would be built. They spent the summer establishing the new workshops and began building Alouette II in September 1963.[40]

Alouette II's design and construction resembled those of its predecessor. Its core consisted of the Alouette I backup satellite that had been built concurrently with Alouette I itself, to be used if the latter was lost. The DRTE engineers modified the backup satellite to include a new probe experiment, and the frequency range of the sounder was increased to better suit the chosen elliptical orbit planned for the ISIS series.[41]

The remaining three satellites would be officially known as ISIS. An ionosonde would serve as the principal experiment on the first two, with additional experiments being solicited from scientific agencies in both Canada and the United States. The plan was to have as many of the experiments as possible on both satellites, so that mutually supportive data of the greatest scientific value might be collected. Primary and secondary payloads for the third ISIS remained unconfirmed, to be determined at a later date.[42]

As with its predecessor, Alouette II's construction remained on schedule, and it was ready for launch in the summer of 1965. The final product carried no less than five experiments, four of which had also been aboard Alouette I; the fifth was designed to measure the sheath, or "halo," of positive ions occurring around the spacecraft's antennas as it passed along the ionosphere. Room for this experiment was created by redesigning and improving the nickel cadmium batteries employed in the original satellite, so that fewer were needed to generate the same amount of power. Removing the extras provided space for the new instrumentation.[43]

On the evening of 28 November 1965, Canada's second satellite was successfully launched from California during its window between 8:40 and 11:30. Riding along with Alouette II was American scientific satellite Explorer XXXI, another payload designed to investigate the upper atmosphere. Alouette reached orbit in a textbook fashion and went into operation almost immediately. Following the launch, Minister of National Defence Paul Hellyer summed up the entire DRTE's feeling when he told public affairs personnel that Alouette II was "an achievement in which all Canadians can be justly proud."[44] Indeed, the DRTE team was more than pleased. The launch of Alouette II marked their second successful satellite project in just three years.

The initial Alouette flights had brought a group of talented professional Canadian defence scientists into the national limelight and underscored what might be possible in the Canadian aerospace industry. DRTE chief superintendent Frank Davies probably stopped comparing the Alouette project to an albatross hung about his neck, and others such as Eldon Warren, John Chapman, Colin Franklin, M.A. MacLean, John Mar, and their colleagues were gaining both reputation and prominence within their community as a result of their efforts. Some of these men, perhaps most notably John Chapman, would move to much higher posts in Canadian scientific administration and government. Some stayed with the DRTE through its transition into the Communications Research Center in later years, whereas others left or retired when it dissolved, moving on to industry or other scientific pursuits. For the moment, however, they were undoubtedly the centre of attention.

Although many delighted in their accomplishment, others, particularly the RCAF, disapproved of the increased focus on Alouette. Desiring a greater role for space technology in Canadian defence, and especially in the RCAF, these critics met the Alouette's success with slight reservation. The program had completely consumed the DRTE's resources and nearly all space technology expertise within the Defence Research Board. Though Alouette's mission did support defence research in some programs, such as defence communications and studies related to ballistic missile defence and ballistic missile re-entry into the atmosphere, the RCAF and other DND planners realized that a DRB agenda was rapidly forming to advocate for the pursuit of pure scientific research at the expense of more defence-oriented technological application. When a DRB proposal to cooperate on the American exploration of Mars appeared in mid-1965, the military felt its concerns greatly justified.

Mars Topside Sounder Project

Every twenty-six months, the orbits of Mars and Earth bring the two planets into relative proximity, making the red planet a bright orange "star" in the night sky. Until the advent of the space age, however, all knowledge of Mars came from astronomical observations alone. Yet once NASA had achieved a degree of sustainable competency in both spaceflight and orbital mechanics, it quickly sought to begin a physical investigation of Earth's closest neighbours.

As NASA's manned spaceflight endeavour was about to enter its second stage with the Gemini Program, its unmanned space science centres prepared for their own first attempts to reach the red planet. Mars and Earth reached closest proximity – known as best opposition – every two and a half years, so NASA planned for an unmanned Mars reconnaissance mission as soon as it was technically feasible.

NASA began its attempt to reach Mars in the early 1960s, achieving incredible technological success on 14-15 July 1965 when spacecraft Mariner IV passed by the planet at a distance of 9,600 kilometres. Mariner IV snapped some of the first blurry close-up pictures of the Martian surface, finally revealing its true nature. Although the scientists back on Earth were a bit disappointed to see that the surface of Mars was moon-like and barren, they were nonetheless impressed that they had finally engineered a successful flyby. The next step was to put a robotic spacecraft into a sustained orbit around Mars.

In 1960 the Jet Propulsion Laboratory in California proposed a new spacecraft program known as Voyager (this should not be confused with the project that eventually sent twin planetary flyby spacecraft to the outer planets in 1977 and 1978). Voyager was initially scheduled for the 1971 Earth-Mars best opposition and would include a number of science experiments designed to provide much more detailed information about the red planet. In June 1965, Homer E. Newell, NASA's associate administrator for space science and applications, made a preliminary announcement of an opportunity for other countries to participate in the 1971 Voyager mission. Wishing to build on their success with Alouette and to undertake a new challenge, the DRTE team responded to the call.

The deadline for proposals was November 1965, only six months away. John Chapman quickly submitted a proposal to NASA for the construction and deployment of a Martian topside sounder satellite based on the Alouette I and II designs. His argument for a topside sounding of the Mars ionosphere was straightforward. He wrote in his proposal that "one condition for the evolution of life as we know it is the presence of an atmospheric shield which reduces solar ultra-violet radiation to levels low enough to permit the formation of the large molecules of life systems ... It is therefore proposed that one of the experiments of a Mars orbiter should be a sweep-frequency topside sounder."[45]

NASA's budget approved in principle the go-ahead for Voyager in 1967, but the death of three American astronauts – Gus Grissom, Roger Chaffee, and Ed White – in the Apollo I pad fire of January 1967 put all plans for a Mars mission on indefinite hold while an investigation into the fire was carried out. Then, in August 1967, the US government took back $500 million from NASA's budget, forcing it to slash several programs to keep the upcoming moon landing missions on schedule. Funding for Voyager ended as a result, and with it, Canada's first attempt at the direct exploration of Mars.

Chapman's Mars topside sounder proposal serves as a critical example of how the DRB had reached a crossroads concerning its overall mandate by 1965. Whatever the scientific and technological merits of the Mars project, it had absolutely no relation to Canadian national defence priorities, which makes one wonder why the expenditure of Canadian defence dollars was even being considered for it. As well, given the possible magnitude of the mission, one might think that the government or even Cabinet would have wanted a greater say in how it unfolded. That no one, especially no one in government, appears to have intervened in this proposal is a tremendous reflection of the autonomy the DRB enjoyed in the mid-1960s to pursue its own research agenda on the international scene. However, it also explains why the diverging interests and priorities of DND's main research and development organization were increasingly incensing the three armed services, especially the air force.

Still, the DRB responded predictably to criticism and concerns from the armed services over such projects. Though the DRB empathized with DND's military requirements, its true cultural interest was the pursuit of pure scientific research, and thus it fought hard to maintain its own agenda regarding scientific satellite research and development. Yet, with political support behind them, or perhaps more accurately, a lack of political interest in obstructing their activities, the DRB senior management no doubt felt that the RCAF or even DND as a whole could do little to alter their chosen course. Interestingly enough, as the decade advanced, it was Cabinet, not DND, that finally forced the DRB to show its hand regarding the nature of its agenda, a process that ultimately led to the demise of the DRB itself.

ISIS: Clipped Wings

Work on the remaining ISIS series satellites began in March 1964.[46] By this time, the Topside Sounder Working Group (TSWG) had resolved its terms of reference and was well into planning and coordinating the ISIS mission profile. J.E. Jackson of NASA chaired the group until 1974, when T.R. Hartz of the Communications Research Center replaced him.[47] The task of the TSWG was to coordinate all planning and scientific aspects of the mission, organize meetings between parties associated with the project, and deconflict scientific and engineering issues as they arose. Although not specifically labelled as such, the TSWG was the systems engineering and project management oversight for the Alouette-ISIS project. To a large degree, the success of the venture depended upon this group.

For the DRB, the technological challenges of the ISIS project were greater than those of Alouette. The ISIS design was more complex, and the engineers needed more time to build and test the satellites. ISIS-A carried eleven experiments, including a fixed-frequency sounder to accompany its swept-frequency sounder and mixed-mode sounder. It also carried a very low frequency receiver and transmitter test, a cosmic noise receiver, an energetic particle experiment, two Langmuir probes that measured electronic temperature and density, two ion mass spectrometers, an ion probe to measure ion temperature and density, a soft particle spectrometer, and a beacon set at 137 megahertz (MHz).[48]

The DRTE was expected to assume responsibility for overall satellite operations. That meant making facilities available to house the control centre as well as training specialized technicians to perform the work of making the satellite fly. The DRTE satellite operations staff were expected to complete spacecraft checkout procedures after launch, inject ISIS into orbit, monitor its health and status in orbit, and change its oriented and spin-axis to accommodate the onboard experiments as required. Mission controllers were also responsible for maintaining the accurate determination of the satellite position in orbit and predicting its position several weeks in advance.[49] The team would have other responsibilities, such as coordinating and deconflicting all transmissions between the satellites and the ground stations, and collecting and processing scientific data.

In support of these tasks, just the computer technology requirements alone were both daunting and expensive. The DRTE had only just finished building its first computer in 1960 and arguably lacked sophisticated expertise to create the systems architecture needed to sustain ongoing satellite command and control operations. Ultimately, other organizations assisted in the task of computer analysis, as the volume of data collected once the satellites were in orbit became too unwieldy for the DRTE to manage on its own. NASA's Goddard Space Flight Center in Washington assumed a data-processing role for the ISIS missions, helping the DRTE to maintain some control over the twenty-five kilometres of scientific-data-laden magnetic tape received daily from the satellite at the height of its operation.[50]

Though technologically rewarding, the ISIS missions proved daunting for Canadian defence science and technology capabilities. The satellites took longer to design and build, and consumed more resources than any of their predecessors. The adaptation of ground-based techniques to a spacecraft in which weight, space, and power consumption were at a premium demanded ambitious and pioneering talent that was soon stretched thin, both at the DRTE and the civilian industries contracted to support the Canadian missions. As well, the sheer magnitude of controlling and operating satellites on a daily basis far exceeded the resources of the DRTE, and the DRB had literally to canvass its own headquarters and Canadian Forces Headquarters (CFHQ) on a regular basis for additional technical resources and funding to sustain operations.[51] Given the increasingly strained relationship between the DRB and the services, one can only imagine how such requests were received.

These realities, combined with other factors then affecting Canada's space program, certainly had an impact on the schedule of the project. The first satellite in the ISIS series, ISIS-A, was not launched until 30 January 1969. Taking off from the Western Test Range in California at 6:46 a.m., the third Canadian satellite was lifted effortlessly into the morning sky upon its booster and entered flawlessly into its assigned orbit of 88.4° with an apogee of 3,522 kilometres and a perigee of 574 kilometres. Early results from the satellite were promising, and it subsequently performed nominally and acquired good data throughout its life cycle.[52]

The next satellite in the series, ISIS-B, did not leave the launch pad until April 1971. Although it too performed as expected, if not better, by the time it reached orbit the focus of Canada's space program was shifting from pure scientific research to commercial applications. As a result, the third satellite in the series, ISIS-C, originally scheduled for a 1974 launch, was cancelled by the government in the summer of 1969.

The Convergence of Government, Defence, Science, and Technology Cultures

Canadian space activities served as a crossroads where three influential Cold War communities converged. These were government and the civil service, defence, and science and engineering. The first two had a long tradition of converging and at times disagreeing with each other, but the third – in this case the defence scientific community – was a new party on the scene, arriving during the Second World War. The relationships within this triad influenced all space program development in Canada during the Cold War period, and understanding the ideals and attitudes prevalent within the defence research culture during the Alouette-ISIS era is critical to defining the relationships that influenced other developments in Canada's space program. What one discovers

FIGURE 32 The end of an era. Already constrained by budget and behind schedule, ISIS-B lifts off at night on 1 April 1971 from Vandenberg Air Force Base in California. Canada's first defence space program would officially end soon afterward. Communications Research Centre, 71-22103.

is that, interestingly but perhaps not surprisingly, Canada's early Cold War experiences resembled those of its two main allies, the United States and the United Kingdom.

Though the Second World War had brought soldier, scientist, and engineer together, once it ended, most scientists and engineers working within Canadian defence establishments preferred to return to their own research interests rather than directly support the goals of the organizations that employed them. Peter A. Forsyth, the talented defence scientist who served as one of four central figures in the Science Secretariat's 1966 National Space Study, later characterized the DRB and the DRTE as somewhat contemptuous of the defence establishment.[53] Scientists expressed numerous concerns about becoming involved in classified projects that would hinder the sharing or publication of their research, as well as open contempt for the politics that might impose specific tasks or objectives on staffs who wanted the freedom to pursue their own goals. Others still, perhaps, preferred not to become deeply involved in military projects that could open the door to another more destructive world war.[54]

Alouette-ISIS simply entrenched this attitude. In the absence of any national space policy or clearly defined space goals during the 1950s, separate defence and scientific establishments were largely free to follow their own research interests regardless of whether these coincided with any national defence objective. The idea for the S-27 Topside Sounder was born this way. When the DRTE was allowed to pursue it with international partners, its realization merely served to validate the DRB establishment's main intent of supporting projects that might garner wider political and financial support in preference to less promising defence assignments. Though this was not necessarily a flawed or unrealistic business practice – after all, DND did feel that it benefited from the acquisition of space technology – it nonetheless created an undesirable precedent that DND was later unable to influence politically or bring under its control.

Opinion also differed regarding the true purpose of Alouette-ISIS. To government and defence, the rationale for committing to it was grounded in the need to advance Canadian space technology for both security and industry. To the scientist, the focus was pure research and data collection for further studies, so much so that the venture nearly took over the DRTE. In a way, projects such as Alouette-ISIS served to ensure the untimely demise of the DRTE. Its near single focus on the development of satellite-communications-related science ensured its eventual transfer as the renamed Communications Research Center to the DOC in 1968-69.

To the engineer, Alouette-ISIS was about developing technology, but the project ultimately exposed many of the weaknesses inherent in Canada's evolving technology base. This is not to suggest that Canada's satellite-building abilities were weak; all things considered, they were very strong. However, as it struggled to develop advanced electronics and improve miniaturization, the DRTE could not draw much-needed resources from its civilian industry (as the Americans could) to sustain an expensive and long-term scientific program, especially when those expenses overtook all available government funding. Whereas, in the United States, private industry paid much of the research and development cost associated with emerging space technologies, in Canada, DND and the DRB were forced by circumstances to assume most of that cost internally.

There is no question that all these groups felt strongly about their roles in developing Canada's space program and regularly clashed throughout the Alouette-ISIS project, but ultimately the government determined the future course for Canada. Despite its best efforts, DND would lose its battle for space control as it engaged in a larger struggle over unification of the forces, and defence-oriented agendas were eventually dropped from national space projects. In the wake of Ottawa's desire to see commercial applications assume more prominence, even John Chapman, the influential coordinator of the Alouette-ISIS project, could not sustain further political support for a pure space science agenda. Despite the changes in political direction, however, Alouette-ISIS marked the convergence of these three cultures and remains one of Canada's most remarkable technological achievements of the mid-twentieth century.

5
Canada's Militarization and Weaponization of Space

Canada's position has always been against the weaponization of space and we will maintain that position.

– Minister of Foreign Affairs Bill Graham, February 2002

Despite the statement above, Canada cannot claim a legacy of opposing the weaponization of space. During the first fifteen years of its space program, it did little if anything to promote exclusively peaceful uses of outer space; instead, like its allies, it realized that its own protection depended upon the military domination of space. During the Pearson era, it shirked United Nations initiatives to promote a space sanctuary, and it investigated and pursued roles that contributed to missile defence, space control, and the potential deployment of space-based weapons. In addition, Canada regularly lent its support and expertise to similar American ventures, many of which also contributed to the militarization and weaponization of space. Only toward the end of the 1960s, largely influenced by political and technological change, did Canada withdraw from such overt military programs. The next two chapters examine its Cold War military programs, drawing upon numerous recently declassified sources that reveal a history far removed from that currently portrayed or politically endorsed.

Military Conflict and Outer Space

At the beginning of the twentieth century, military conflict had evolved from ground to air, and from the dawn of the Cold War, the United States and the Soviet Union expected that it would transcend into space by the end of the century. Immediately after the Second World War, the two new superpowers devoted considerable attention and resources to preparing for what seemed to be inevitable – a war in space. Though today notions of torpedo-laden spaceships securing the stars are the stuff of science fiction, in the early 1950s, it was anticipated that space would simply become yet another medium to defend against and to dominate.

As the naval powers had contemplated the globe, so did the first atomic powers contemplate space. There were many questions and concerns about this new ocean. Speaking of the atomic powers, political scientist Hedley Bull pointed out in his 1961 work *The Control of the Arms Race* that "their existing legal and political conceptions do not cover it, and their experience provides them only with analogies." He added, "They can have little notion of the problems to which it will give rise, or of the political, strategic, and economic importance it will have for them."[1] The transfer of military capability and power into space gave rise to new terminology – the words "militarization" and, later, "weaponization" – which would have significant political and legal ramifications, both domestically and internationally.

The initial concepts were relatively straightforward. The phrase "militarization of outer space" simply referred to the employment of artificial satellites that were supportive of or ancillary to terrestrial national security and/or the projection of military power. The "weaponization of space" referred to the employment of weapon systems in or through space (such as ICBMs, which travel through space en route to their targets) or more obviously to the transformation of space itself into a theatre of military operations.[2]

Arguably, the militarization of space occurred immediately after the first satellite launches in 1957 and 1958, when American and Soviet militaries began deploying their surveillance and reconnaissance assets. The weaponization of space was realized earlier in the 1950s, after the first ICBM and anti-ICBM tests and deployments. The actual transformation of space into a theatre of operations came later, when the Soviet Union began testing and deploying an anti-satellite system in 1963, followed by its deployment of a fractional orbit bombardment system in 1967.[3] In response to the construction of new Canada-US early warning missile-tracking systems in the Far North, the bombardment system was designed to get nuclear warheads into the United States by using a southern trajectory. Launched into an orbit of 160 kilometres and then decelerated to allow re-entry, the nuclear warheads would hit their targets before the completion of the first orbital revolution. In time, the Canadians and Americans countered with their own anti-satellite programs, and later still, the Americans pursued a Strategic Defence Initiative, which proposed many orbital space weapon platforms. Needless to say, national security trumped other concerns, and both camps rapidly ignored the concept of treating space as a peaceful sanctuary.

Space: Canada's Next Battleground?

Like its adversaries and its allies, Canada planned, organized, and directed its early rocket and space development largely in support of its Cold War security apparatus. It was no less ambitious or aware of the potential that access to space could provide; nor was it ignorant of the incredible threat that it posed.

Accordingly, Ottawa continuously sought ways to maximize its access to and control of space, but given its limited resources, this meant investing considerably in both indigenous and bilateral defence space programs.

Soon after the launch of Sputnik, Canadian strategic thinking on outer space by analysts such as R.J. Sutherland and others within DND started to push beyond purely scientific investigation toward military applications and by default to civilian applications as well. Interestingly, the defence pundits alone did not lead this shift; rather, other government departments, especially those concerned with the identification, advancement, and diffusion of the technological innovation derived from rocketry and space, also came into play. As well, even academia assumed major roles at times, but despite this external input of ideas, DND continued to evolve and fund the majority of its own early post-war research and development.

Gaining a clear understanding of how the main agencies interacted during this period remains difficult. The Science Council of Canada suggested at the time that technological underdevelopment was a leading cause of arrested national economic growth, a situation for which all agencies involved in this field were in part responsible. Yet verifying this charge is not easy, especially because most literature on the politico-economic implications of technological development in the post-war era consists of 1970s government studies and analyses produced in response to growing government concerns over the deindustrialization of Canada, not independent assessments of the situation by other interests or observers. Still, though it is beyond the scope of this study to detail the history and historical discourse surrounding the last decade of Canada's post-war economic expansion, it is important to note that technological development was considered a strategic national interest and that all government agencies, regardless of their mandate, were expected to focus on it. For DND, this meant playing an important role in potentially shaping future technological and economic change, by linking national defence with national welfare.[4]

Though the DRB continued as the primary agency for missile and space research and development throughout the late 1950s, the RCAF began taking direct control of operations as early as July 1957, when the recently elected Diefenbaker government ratified its decision to share the responsibility for North America's air defence with the United States, which in turn led to the creation of NORAD. Originally tasked with providing strategic early warning as well as coordinating armed defence and response to attack, NORAD soon assumed other roles such as space analysis, satellite tracking, and, later, anti-ballistic missile defence. For the RCAF, this meant adopting responsibility for and, by default, control of these space-related areas on behalf of DND at home.[5]

Possibly because of either its increasing role in the newly created NORAD or its growing involvement with US-led ballistic missile re-entry research, the RCAF was unique within DND for its technological and operational focus on outer space. The Royal Canadian Navy was largely preoccupied with fleet digitization and automation at this time, and the Canadian Army was immersed in addressing its own combat development in preparation for the nuclear battlefield.[6] Thus, it was logical for the air force to want direct involvement in and control of Canada's fledgling military space program, just as the air forces of its allies were becoming responsible for their respective defence space programs as well.

What the RCAF did not expect was an all-out political battle with the DRB to become the office of primary interest for the development of Canadian military space applications. Nor did it expect to see the day in which the DRB would suggest that its own mandate and primary function was pure scientific investigation, not defence research and development. Yet these competing philosophies existed, and the political infighting they created plagued military space planners until the battle for control of government space programs was finally lost.

Though the RCAF promoted several advanced missile and space concepts throughout the late 1950s and early 1960s, it should be noted that they were often closely linked to American efforts in that they primarily addressed the cooperative defence of North America against Soviet attack. However, this is not to suggest that Canada simply provided ancillary input to a largely autonomous American program: rather, the RCAF excelled in certain niches. On many occasions, Canada's air force acted not only as the subject matter expert and primary researcher and developer but also as the sole provider of a particular capability. Given this, Canada's lesser physical and financial contributions to the bilateral militarization and weaponization of space in the 1960s should not be taken out of context. As demonstrated later in this chapter, Canada was not reluctant to militarize or weaponize space; in fact, the record indicates that, had more resources been available, it might have fielded a higher-profile effort prior to the end of the 1960s. Instead, national security, fiscal realities, and strategic interests prompted Canada's decision to combine with American-led defence space initiatives rather than attempting a parallel space program of its own.

Anti-Ballistic Missile Defence Projects, 1958-63

Once the threat of Soviet ICBMs to North American security was validated beyond doubt, Canada's military wasted little time in cooperating with its American partners to develop sophisticated and far-reaching early warning radar, anti-ballistic missile defence, and space control assets as quickly as possible. The government had constructed and linked strategic air defence systems across northern Canada; now it also provided missile launch facilities in the

North and cooperated with the American military on the testing and evaluation of various ballistic missile defence systems. With this, it accepted an informal if not officially recognized role in the weaponization of space.

The joint Canada-US winter test-firing program started at Fort Churchill in 1955, with the objective of investigating the effects of Arctic conditions on the American Nike Ajax missile system. At the time, this missile system was being adapted to deliver a nuclear weapon against inbound Soviet air-breathing and unmanned threats from the Far North, and thus it needed to be evaluated in the Arctic environment. The Canadian military, which would probably operate those defences deployed on Canadian soil, took an active role in the test and evaluation phase. The program concluded in late 1957 as preparations were under way for continued testing on other American-built missiles.

Canadian contributions to ballistic missile defence research and development evolved quickly. In June 1958, the DRB reported that "the ballistic missile defence program has now been in operation for about two years with some elements of it being about four years old. The major contributions are in the three establishments DRTE, CARDE, and ORG [Operations Research Group]. DRTE is conducting research in the propagation, background noise, communication, and reflective area fields. CARDE is working in systems analysis for active systems, infrared propulsion, and aerophysics fields. ORG has undertaken analysis and evaluation of the systems for ICBM defence."[7]

Clearly, with three of its main research facilities engaged in ICBM-related research, the DRB considered the development of Canadian ballistic missile defence capabilities a priority, but this should come as no surprise given the history of the period. By 1958 the Soviets had not only detonated a 1.6-megaton thermonuclear bomb (capable of up to 6.0 megatons), they had also discovered how to get it into space and thus potentially to anywhere in the world.[8] The pressure was on to defeat such an attack before it could occur.

Following the completion of the IGY program in 1958, but prior to the initiation of the Black Brant project, discussions were held on the coordination of a joint Canada-US guided missile range at Fort Churchill. The US Department of the Army represented American interests, whereas the Canadian Army spoke for Canada at these meetings on 14 May 1959. An Ad Hoc Joint Operational Coordination Group was formed and met for the first time in Washington on 1 July 1959; a second meeting followed in Ottawa on 25 August. Plans for a number of launches were coordinated and, subject to government approvals, would proceed early the following year.[9]

Development of the joint program continued throughout the winter of 1958-59, subsequently leading to the testing of the newly developed Nike-Hercules and Lacrosse missiles, launching seven of the former and eight of the latter. The

FIGURE 33 Canada-US anti-ballistic missile defence projects and data eventually led to the development of sophisticated integrated early warning systems such as the Semi-Automatic Ground Environment (SAGE) Direction Center. Here, two RCAF technicians operate an early warning console in the 1960s. National Defence Imagery Library, PCN-4714.

following winter, in 1960, US Hawk and Redstone missile systems were added, the latter of which was converted in early 1961 to carry the first American astronauts into space.

The impact of expanding missile-testing activities was felt in the barren plains of northern Manitoba. By 1960 the Canadian Army had nearly 100 people assigned to the American range at Fort Churchill, and the US Army had just over 550 people at the facility. In addition, the USAF had nearly 180 officers and men stationed there, bringing the total military cadre, including support and security personnel, to approximately 1,000 soldiers and airmen. Again, the contribution in human resources underscored the degree of effort being made to find a means of deterring strategic attacks from the Soviet Union during this period.

While additional tests were carried out on the Sergeant missile and the British Thunderbird missile during the winter of 1961-62, the RCAF began to expand

FIGURE 34 Members of the Royal Canadian Artillery attend a familiarization course on the Nike-Hercules missile in the United States during the late 1950s. National Defence Imagery Library, Z-8435-1.

its own ICBM research into other areas, such as ballistic re-entry characteristics. Its expertise soon reached a level where even the USAF was requesting Canadian aid in advancing its own emerging anti-ICBM research projects. The RCAF gladly responded to such requests, and between 1960 and 1965, the Central Experimental and Proving Establishment (CEPE) cooperated with USAF missile re-entry research under the code name Project DEFENDER. Heavily funded by the US Advanced Research Projects Agency (ARPA), this involved not only the RCAF but also the DRB and even CARDE at various times.

The scope of Project DEFENDER was considerable, and certain aspects of it remain classified and therefore beyond the reach of historical investigation. Although giving a complete account of the project is impossible, some parts of

FIGURE 35 A Nike-Hercules missile is launched from the Canada-US joint test facility at Fort Churchill, Manitoba, 1960. National Defence Imagery Library, Z-8464-2.

it are now declassified and better known to us. Two such examples, Operation LOOKOUT and Operation BLIND TWINKLER, are explained in greater detail here to provide some context for this highly important Canada-US space defence cooperation at the height of the Cold War and the Cuban Missile Crisis.

Operation LOOKOUT got under way in January 1960. Thirty RCAF personnel under the command of Flight Leader Murray Sweetman and four DRB scientists led by Guy Giroux were sent to Ascension Island, a remote speck of rock approximately five thousand miles downrange from the missile launch sites at Cape Canaveral, Florida.[10] The aim of LOOKOUT was to measure the infrared radiation characteristics of ballistic missiles launched from Cape Canaveral. In particular, the DRB scientists were trying to define the radar characteristics of an ICBM nose cone upon re-entry, so that early warning radars might be calibrated accordingly to detect incoming Soviet missiles.

Mounting wing pods with specially designed detection equipment on two CEPE CF-100 Canuck fighters (affectionately known as Clunks to their pilots), the RCAF crews flew two series of tracing missions to capture data from a wide range of missiles. The two fighters were required to intercept and photograph the inbound supersonic warheads before they landed in the ocean, demanding considerable skill from the pilots, who were unfamiliar with intercepting such swift-moving targets. Yet their skills proved worthy: an estimated 90 percent of the data recorded during the first series of tests proved valuable to further research.[11] The DRB chairman, A.H. Zimmerman, even remarked, "the flying and navigation skills exhibited by the personnel concerned should be a matter of pride to all in the RCAF."[12] Likewise, the chief of the air staff, Air Marshal Hugh Campbell, congratulated his men on a job well done.

The Americans were also suitably impressed with Canada's efforts. In mid-1961, as the first phase of LOOKOUT drew to a close, ARPA representatives asked whether the DRB was interested in continuing its research. The Americans were planning to test their second-generation missiles and wanted to collect data like those taken during the first phase of Operation LOOKOUT. The new missiles used improved fuels, and their radiation characteristics would differ from those of their predecessors. In addition, ARPA wished to capture this data for other research as well.[13]

Having enjoyed considerable success during early 1961, Canada responded favourably, and phases two and three of Operation LOOKOUT ran through to September 1963. These phases were focused at the opposite end of the missile trajectory where the Ascension Island crews were following, measuring the high-altitude effects of missiles leaving the atmosphere and collecting data during what was then called the sustainer phase of ICBM flight.[14] Again, the

FIGURE 36 A CF-100 arrives at the airbase on Ascension Island after collecting US ballistic missile re-entry data as part of Project LOOKOUT. National Defence Imagery Library, PL-120190.

RCAF crews employed CF-100s flown from bases at Grand Bahamas Island and Eleuthera, placing them downrange from the Cape Canaveral launch sites.[15]

Similar in scope to LOOKOUT, Operation BLIND TWINKLER was aimed at establishing the feasibility of using reflected sunlight techniques to detect ballistic missiles in mid-course. In June 1962, twenty-two RCAF crew and five civilian scientists from CARDE reported to the USAF base in Thule, Greenland, to fly a series of high-altitude missions focused on capturing data related to the unusual infrared background effects that were likely to be detected during the "sudden warming" that occurs at high altitudes in the Arctic each spring.[16] Twenty flights above forty thousand feet were carried out from Thule and Fort Churchill during the summer and fall of 1962, with additional flights from Thule during the first half of 1963 to determine whether the magnitude of background signals against which ICBMs were typically detected had altered.[17] Led by Flight Lieutenant J.F. Dyer, the operation was another RCAF success, and it provided much valuable information to DRB and ARPA scientists working on the ballistic missile defence problem.

The RCAF and Space Indoctrination

On 3 April 1959, the RCAF chief of operational requirements presented a study on the military potential of space to the Air Council at Air Force Headquarters.[18] The study emphasized three main points of interest. First, a space-based missile carrier was unlikely to replace the ICBM as the primary form of deterrence unless the Soviet Union developed an effective defence against ballistic missiles. Second, there appeared to be little use for military forces in space, but third, satellites would very probably prove useful for a myriad of defence functions ranging from intelligence gathering to weather forecasting.

A remarkably accurate prediction, the presentation sparked considerable interest and discussion in the Air Council, prompting it to generate a plan to explore the potential application of military space programs within the Canadian forces. Furthermore, as the air force was the obvious lead agent for such programs, recommendations were made to investigate options for increasing the base of space knowledge within its ranks. Senior air force commanders were directed to explore the feasibility of establishing educational programs to afford RCAF personnel a good knowledge of the developments and problems in space technology, and to study the possibilities for integrating select RCAF members into American missile and space programs at all levels. Furthermore, these educational programs were to include indoctrination in the field of nuclear warfare, as both ICBMs and spacecraft were expected to mount these lethal warheads. Space and nuclear briefings were to be delivered to all air force officers of the rank of group captain and above.[19]

Plans for the space indoctrination of the RCAF were executed quickly. Within weeks of the April 1959 Air Council meeting, Air Force Headquarters had developed a space indoctrination course consisting of a series of lectures covering the physics of the solar system; the characteristics of rockets, ballistic missiles, and space vehicles; the potential uses of space vehicles; and a review of Canadian work to date in the various fields covered in the course. In 1960 the course was presented at Air Force Headquarters, Air Material Command, Air Defence Command, Training Command, and 1 Air Division. Whereas the commands gave the course only to their headquarters personnel, 1 Air Division presented it to all wings and included presentations to both commissioned and noncommissioned officers.[20]

At the same time, the air force chief of operational requirements and K.J. Radford, the director of systems evaluation, organized a study program to evaluate how future RCAF operations might prove useful to several American space initiatives. Collectively known as the Advanced Technology Evaluation Program, the series of reports it generated provided the basis for determining future requirements and the extent of RCAF participation in applications of advanced technology; it also increased the general level of understanding of the impact of these changes.[21]

Initial studies completed during 1960 and 1961 under the program identified four main areas of potential RCAF participation in the rapidly evolving American space program. These were space surveillance, satellite communications systems, navigation satellites, and ballistic missile defence, particularly the missile detection alarm system (the MIDAS early warning program). Additionally, the RCAF expressed interest in the American man-in-space program and in integrating one or two Canadian officers into it, though admitting that this was unlikely to prove of great value to its specific plans.[22]

To some degree, Canada had already engaged in these four areas. Since the late 1950s, the RCAF had actively participated in operations such as LOOKOUT and BLIND TWINKLER, the results of which would prove useful to the MIDAS program. Canada had also advanced in space surveillance, and the recently installed Baker-Nunn satellite-tracking camera station at Cold Lake was proving of critical value to early American collection of Soviet satellite intelligence.

Likewise, senior Canadian security and intelligence officials were made privy to the considerable results obtained from the highly classified American Corona imagery satellite program, adding to Canada's knowledge of the potential in this area. John Starnes, a senior Canadian security and intelligence official who later wrote a significant memoir of his career, even had the privilege of visiting the National Photographic Interpretation Center in its well-concealed and inconspicuous rundown building on the outskirts of Washington, DC, some time during 1961 or 1962.[23]

Finally, though neither the RCAF nor other Canadian agencies had made dedicated advances in satellite navigation, the DRTE was heavily involved with preparations for the launch of Alouette I. This was perhaps the only sector in which Canada had yet to demonstrate a serious effort.

Canadian Servicemen in American Space Programs

As the space indoctrination course got under way in 1960, the RCAF initiated further plans to integrate members into the evolving American missile and space architecture. Discussions through its air liaison officer in Washington, DC, with the American armed services verified the potential for this. At first, progress was slow, as the United States was deeply engaged in many programs and was unsure whether the addition of foreign officers would help or hinder American efforts in space. However, USAF general Curtis Lemay took a personal interest in the Canadian offer, especially in receiving RCAF officers with science and engineering backgrounds, and as a result a number of postings were identified before the end of the year. By December 1960, negotiations were in motion for the integration of Canadian officers into selected American programs.[24]

In 1961 the RCAF selected its first twelve officers with technical backgrounds to participate in a wide variety of American space programs ranging from launch systems to manned space flight. Officers in this venture, which defence documentation referred to as the Space Indoctrination Program (SIP), were normally assigned for a period of three to four years, though their tour of duty could occasionally be shorter.[25] Most came under the command of the USAF Headquarters Space Systems Division in Los Angeles, though they worked at various facilities throughout the United States.[26] The Space Systems Division was tasked with the planning, programming, procurement, development, and management of dozens of space projects and systems, and it acted as a primary centre for capabilities and future systems research, including weapons systems concepts and development.

From the first group of RCAF officers, Squadron Leaders John Webster and Allan Pickering were initially assigned to the Mariner and later to the Atlas-Agena Program Office. Flight Lieutenant Andy Thoma worked as a project officer in the Engineering Division of the Standard Launch Vehicle II Directorate.[27] Squadron Leader Jack Henry, stationed in the Aeronautical Systems Division at Wright Paterson Air Force Base, did extensive work alongside Flight Leader J.H. Lathey, who was engaged on the X-20 Dyna-Soar Program. One of the most advanced space programs in the United States, it was eventually cancelled in December 1963.[28]

Squadron Leader Richard "Bud" White had perhaps the most interesting, if at times peculiar, assignment to the Mercury and Gemini Launch Vehicle Directorate. Positioned at the heart of the space race, this office was responsible for getting American astronauts into orbit and eventually to the moon, preferably ahead of the Soviets. "I spent most of my first year seconded to the Mercury Program Office," White later recalled, "traveling with the program director, accepting launch vehicles at General Dynamics Astronautics, San Diego, and participating in the final two launches of the Mercury Program at Cape Canaveral. My basic task had been to bring the lessons learned on Mercury back to Gemini [a two-man space capsule]."[29]

White's duties did not end there. With the ramping up of the Gemini Program, he noted, "As the GLV [Gemini launch vehicle] Pilot Safety Operations Officer, I became responsible for all acceptance and flight safety and pre-launch review board operations, and also for propellants, loading, and engine operations at Cape Canaveral."[30] This was a considerable degree of responsibility and trust

FIGURE 37 A Canadian myth debunked. RCAF exchange officers – such as Captain C.S. Lines and Major B.C. Dimock, seen here working in the USAF Space and Missile Systems Organization Headquarters – were routinely involved in all aspects of American defence space projects during the Cold War. Unattributed RCAF photo, *Sentinel* magazine, September 1969.

given to a young officer who shocked his NASA colleagues on his first day at work when he arrived wearing an RCAF uniform.

As noted above, these "foreign" exchange officers were afforded a substantial level of responsibility, and their success in their assignments demonstrates the quality of RCAF officers at that time. At the Air Force Flight Dynamics Laboratory, J.H. Lathey made a name for himself through his contributions to the guidance and control and "fly-by-wire" systems employed in the X-20 Dyna-Soar. Likewise, Squadron Leader MacFarlane did considerable work on hypersonic ramjet research and development at Arnold Engineering Development Center, Tennessee, and Flight Lieutenant T.M. Harris worked on spacecraft re-entry at the Flight Dynamics Laboratory.[31]

Canadian air force members were also assigned to other US offices such as the 496L System Program Office in Massachusetts, which dealt with space object detection and tracking, or to the USAF Space and Missile Systems Organization Headquarters. By 1967 as many as thirty RCAF officers had served in American space and missile programs directly under the SIP, though their actual number, not counting those involved in NORAD Air Defence Command and other related exchanges, is estimated to be much higher.[32]

Of the seventeen SIP officers who had completed their tours by 1967, four had been released from the Canadian forces, two were assigned to the Central Experimental and Proving Establishment (CEPE), four were assigned to CFHQ, two were attending courses, and the remaining five were scattered across the Canadian military, though the whereabouts of one were reported as "unknown."[33] The SIP continued until at least 1970, after which point, records of exchange duties within American space and missile programs remain classified.

Space Surveillance and Reconnaissance

In the West, the catalyst for designing and launching an earth satellite was based on the critical requirement for advanced strategic surveillance and reconnaissance of the Soviet Union. The United States devoted considerable resources to this enterprise, committing to a highly classified satellite reconnaissance program very early in the space race. Later called Corona, the project was operational before 1960 and ultimately included 145 satellites that photographed nearly 487 million square nautical miles of foreign territory.[34] Corona was shrouded in such secrecy that only a few foreigners, including a handful of very senior Canadian security and government officials, knew of its imagery products let alone the assets that collected them.

Unclassified archival research, documentation, and personal claims give little indication of how extensively informed senior Canadian officials were regarding early Cold War American secret satellite systems such as Corona. As recounted above, John Starnes mentioned visiting the National Photographic Interpretation Center in 1961 or 1962, but he did not indicate whether he alone received this privilege. It is probable that senior members of joint intelligence communities and certain members of DND and the military had some awareness of American photo-reconnaissance capabilities, but only further investigation and declassification of archives will resolve this issue.

Key sources on Canada's role in strategic surveillance and reconnaissance have yet to be declassified, and the limited primary sources that are available provide precious little information. Even secondary literature on the subject is sparse, with only a few articles discussing obsolete methods appearing in some journals over the decades.[35] Not until the late 1970s, when Ottawa was initiating its own space-based remote sensing program, was the issue of space-based surveillance and reconnaissance even discussed in the Canadian context. Like most of Canada's missile and space history to date, this subject has received little attention, and only more scholarship and the passage of time will alter the situation.

Still, surveillance and reconnaissance were not restricted solely to looking down at the earth. As the Soviet satellite population continued to grow rapidly, it became obvious to Canadian and American intelligence analysts that the USSR was probably using space for strategic surveillance and reconnaissance. Given this, there was a requirement to look up – to identify and, if possible, to track foreign satellites passing overhead to determine what threat they might pose to North American security.

Shortly after the October 1957 launch of Sputnik, the USAF Cambridge Research Laboratory initiated an in-house project to obtain orbital elements on Soviet spacecraft. At first, the project was little more than a quasi-scientific effort, collecting data from a diverse range of telemetry readout stations, scientific radars, and even amateur astronomers. However, when the ever-increasing number of Soviet satellites and spacecraft became a matter of national security, steps were taken to formalize a space detection and tracking system (SPADATS) under the aegis of the newly created NORAD.[36]

In early 1959, the USAF established the inconspicuously named 496L System Program Office at Hanscom Field, Bedford, Massachusetts, and gave it the responsibility of developing a military satellite-tracking system.[37] Initially, resources were extremely limited, and sufficient financial support to undertake expensive sensor developments was not available until 1961. In June of that year, the Hanscom data-processing centre was transferred to Cheyenne Mountain at Colorado Springs, where Air Defence Command assumed responsibility for its operations, and the system was renamed SPACETRACK. Later, it was integrated into the NORAD SPADATS system and controlled through the newly created Space Defense Center, also located within Cheyenne Mountain. SPADATS was then expanded to include the US Navy Space Surveillance System and later Canadian contributions as they progressed. Except for the data-processing centre, the 496L Office remained at Bedford, where it continued to improve the overall system architecture.[38] Canada formally joined the American SPADATS effort in late 1960, providing personnel and equipment in Canada and the United States, including its own Satellite Identification Tracking Unit (SITU) at RCAF Station Cold Lake, Alberta, and liaison officers to the 496L Office at Hanscom in late 1961.[39]

At Cold Lake, the SITU employed a Baker-Nunn optical satellite-tracking camera to track designated targets. This type of camera followed its target by taking advantage of the sunlight that it reflected, but the sensor could operate only in darkness while the target was reflecting sunlight and with a suitable angle of reflection – the "look-angle" – between the two. The Space Defense Center at Cheyenne Mountain then informed the SITU camera crews where a satellite would be at a certain time so that the crew could photograph it at a good look-angle against a background of dark sky. Once the film was developed, a star atlas transparency marked with a grid reference system was superimposed

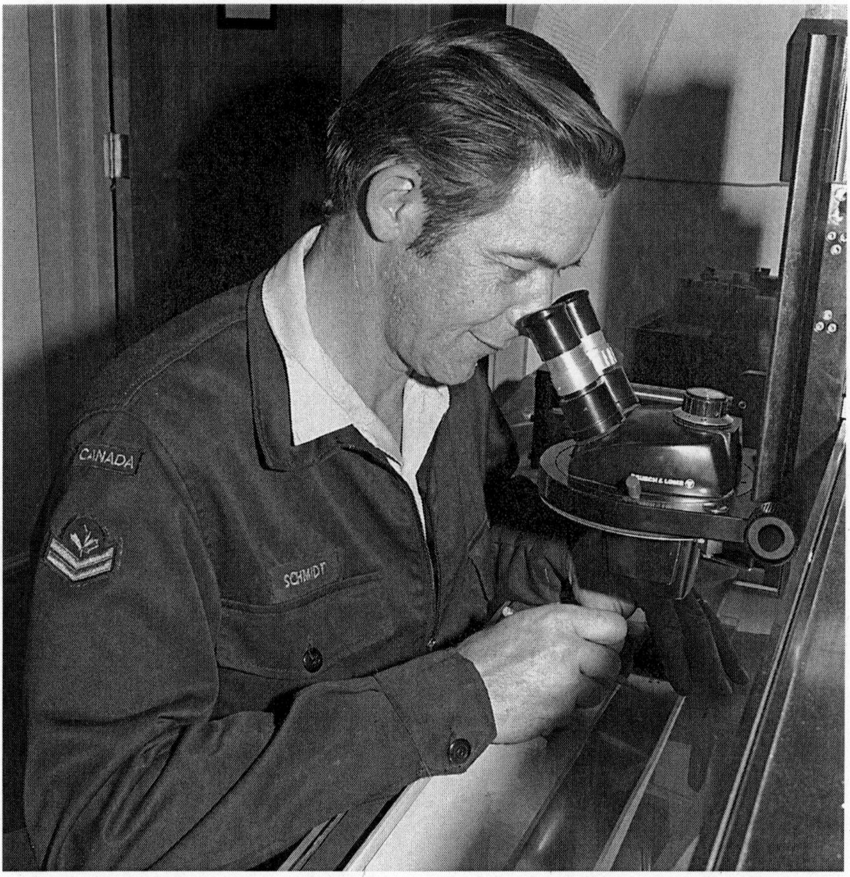

FIGURE 38 Photographic technician Master Corporal Ed Schmidt analyzes film from the Baker-Nunn camera. Data derived from these intelligence sources were critical to early Cold War defence decision making. National Defence Imagery Library, IH78-247.

over the photograph and oriented to the stars on the film. With this done, it was possible to read off the right ascension and declination for any object shown on the film, thus identifying its position in the sky. For high earth and geosynchronous orbit satellites, this method was particularly effective, allowing for accuracy to 30/3,600ths of a degree or 30 seconds of arc. It was claimed that the process could be further refined if needed, increasing the accuracy to 2 or 3 seconds of arc.[40]

The Baker-Nunn camera had a five-degree by thirty-degree field of view, and the length of exposure was optional, allowing the crew some flexibility in tracking and photographing both stationary and fast-moving targets. Once the film

FIGURE 39 Members of the RCAF Satellite Identification Tracking Unit (SITU) at Cold Lake, Alberta, prepare a Baker-Nunn satellite-tracking camera for operations. National Defence Imagery Library, NB65-841-1.

was processed and reduced, the satellite position data were sent back to the Space Defense Center where it was processed by a computer and entered into a catalogue. The catalogue grew quickly, and the computers worked tirelessly as dozens of space objects multiplied by hundreds and soon by thousands.

From the outset, it was clear that the technical analysis of tracking objects in space required a huge degree of computational power, at a level not yet developed in Canada. Some accounts have suggested that there were only four computers in Canada in 1960: a British-made FERUT at the University of Toronto, an American-made Computer Research Corporation CRC102A at RCAF Station Cold Lake, an American-built ALWAC III at CARDE, Valcartier, and the newly designed DRTE computer, nicknamed Girtie, in Ottawa. It was the CRC102A at Cold Lake, combined with the Baker-Nunn camera operated by the SITU, that constituted the Canadian military's main contribution to SPADATS during the 1960s.[41]

The Co-Orbital Satellite Intercept Evaluation (COSINE) Project

The increased challenge of defending against a Soviet ballistic missile attack during the 1960s focused the early efforts of RCAF missile and space research. As the SPADATS program had revealed, Soviet reconnaissance of North America from space was probably expanding, given the rapid increase in its unmanned launches throughout the last five years. Despite the drought of available unclassified information, there is little doubt that Canadian strategic intelligence estimates of the time reported that Soviet reconnaissance threatened Canadian and American security.[42] Accordingly, the RCAF proposed to complement its evolving satellite-tracking capacity with an anti-satellite intercept deterrent capability. With this combination, the military could easily claim a salient contribution to the strategic defence of North America.

In 1959 the RCAF Directorate of Systems Evaluation began a series of studies and computer simulations of the ground and space problems involved when one space vehicle rendezvoused with another in orbit.[43] Known as the Co-Orbital Satellite Intercept Evaluation (COSINE), the studies dealt with the co-orbital rendezvous problem, a critical piece of the space puzzle that affected everything from satellite launch, to anti-satellite weapons, to missile defence, and even to landing a man on the moon. COSINE's ultimate purpose was to determine the most feasible means of moving from the ground to a target vehicle and to develop the guidance procedures necessary for closing the gap between the two vehicles to a predetermined amount. As well, it investigated probability estimates of mission success, taking ground environment, launch vehicle, and rendezvous system vehicle errors into account.[44]

Throughout the modelling and simulation, DRB scientists assumed that the target vehicle, presumably a Soviet satellite, would neither obstruct nor assist the intercept. Within these parameters, the studies proved successful and indicated that rendezvous intercept in orbit from a ground-based launch would be feasible. As far as equipment was concerned, the studies noted the requirement for a vehicle, such as a missile or spacecraft, with a mounted automatic terminal homing guidance sensor for achieving and maintaining the desired rendezvous conditions.

Preliminary reports to the recently appointed chief of the air staff, Air Marshal Clarence Rupert Dunlap, were favourably received. In June 1963, he ordered the RCAF Directorate of Advanced Engineering and Development (DAED) to extend the COSINE studies into a full research and development program, and he arranged that funding be set aside for it. Interestingly, there appears to be no record that Dunlap considered such an assignment to lie within the realm of DRB responsibilities, but planners at the DAED and other air force departments knew that DND's primary research establishment would need to be consulted and sold on the idea before it executed any new initiative that would result in competition for precious technical and engineering assets. Despite the apparent ease of Dunlap's decision, much work remained before the RCAF could see it realized.

The RCAF Space Defence Program

The RCAF-advocated Space Defence Program, sometimes referred to in government documentation as the Space Development Program, evolved from the results of the well-received COSINE project. In June 1963, after being impressed by the success of initial COSINE reports, the Air Council approved in principle the recommendations of a requirements study, which suggested that future RCAF military space projects should include the development of co-orbital rendezvous techniques.[45] Subsequently, $2.2 million was committed to these investigations, to consist of advanced research and development of terminal homing guidance sensors for satellite interceptors. Essentially, the proposed system resembled the exoatmospheric kill vehicle developed by the United States in support of its current National Missile Defence Program – except that the RCAF was investigating this means of missile defence over half a century earlier. After the Treasury Board approved funding in principle in July 1963, the DAED was tasked in August with initiating the Space Defence Program, which would carry a design through the hardware and flight test stages.[46]

Yet the idea continued to evolve after the original parameters were set. By December 1963, the DAED had advanced the scope of the co-orbital interceptor concept to include inspection sensors, non-nuclear kill mechanisms, and negation capabilities. These were considered useful because they increased the overall ability of the technology and because Canada would be exploring an area that the United States had not yet investigated. Although it was later learned that such was not the case, the US Department of Defense expressed its support of and interest in Canadian studies, tests, and evaluations in this field.[47] RCAF senior personnel met with their American counterparts in the US Directorate of Defence Research and Engineering (DDR and E) at the Pentagon in March 1964. Their discussions of options for cooperation and collaboration led Harold Brown, the DDR and E director, to later provide testimony to the United States Senate Committee on Aeronautical and Space Sciences supporting Canadian research into co-orbital rendezvous techniques.[48]

American interest in Canadian orbital rendezvous studies should not be understated. In early 1964, the United States had yet to succeed with orbital rendezvous and was reaching a critical point in its military and civilian aerospace programs that ultimately required mastery of the skill of meeting in space. On the military side, growing allied concerns over Soviet ballistic missile capabilities had already dictated the likelihood that the only plausible defence against an inbound nuclear missile was to destroy it with another missile. In such a situation, the inbound missile must be hit; thus, if any uncertainty existed on this point, the United States would probably resort to the use of a nuclear weapon as it had greater destructive power than a conventional one.

On the civilian side, the American plan for landing a man on the moon depended on the ability of two spacecraft to meet, not only in Earth orbit, but eventually in lunar orbit as well. In 1964 the United States had launched the first of two unmanned flights in Project Gemini – the second stage of its massive attempt to reach the moon – and an additional ten manned flights were scheduled to be completed no later than the autumn of 1966. Of these ten flights, no less than six were primarily devoted to developing and perfecting rendezvous and docking procedures.[49] Having two multi-ton objects meet and potentially grab onto each other while travelling in the vastness of space at 27,369 kilometres per hour more than two hundred kilometres above the earth was no small feat, but it was absolutely necessary if both military and civilian programs were to prevail.

Project Gemini devoted a huge effort to orbital rendezvous, but one should remember that it involved a manned spacecraft encountering a cooperative target sporting a locating transponder. The RCAF project entailed the more difficult task of the orbital intercept of a non-cooperating target.[50] During the RCAF visit to the Pentagon in 1964, the Americans had admitted that they had not yet initiated a serious study into non-cooperative orbital interception, but they were interested in supporting and cooperating with Canadian efforts in this field.[51] The Pentagon had suggested that Canada might lead the way rather than simply being an adjunct to a larger American project. As of 1964, the United States had only a few ballistic intercept capabilities, consisting primarily of short-range Nike-Zeus and Thor-Agena missiles tipped with nuclear warheads. However, though the Nike-Zeus could be launched from the Canadian Far North, the Thor-Agena could not. Thus, the Americans wished to block Soviet access to space and were naturally interested in an RCAF proposal that might achieve this end.

With this and other recommendations, the Space Defence Program had started to take shape by mid-1964, at least on paper. The next step was to identify the agencies that would execute the program. In the United States, each of the military services conducted its own defence research, but such was not the case in Canada: the Air Council knew that its own arm of the military could not design and execute the program without external involvement. Except for the initial concept phase, all efforts had to come from the defence scientific community – namely, the DRB and its suborganizations.

On 14 May 1964, after further planning, all parties involved in the RCAF Space Defence Program met at the Ottawa offices of the Defence Research Board. The team representing the air force consisted of Squadron Leader N.B. Flavin and J.H. Crysdale (Squadron Leader J.C. Uhthoff was designated Flavin's alternate in the event of his absence). The agenda was, if possible, to authorize the space defence program proposal as tabled and get the Canadian forces officially involved in the business of military space.

The RCAF program, which remained focused on the development of spaceborne equipment for satellite interception, was divided into three smaller projects.[52] The first was the research and development of terminal homing guidance technologies. Working from the assumption that very small miss distances were allowable in order to achieve close inspection or a non-nuclear "hit" on a target satellite or spacecraft, the project contemplated investigating the employment of a guidance subsystem based on either passive or active spectral (radar or optical) sensor techniques. The proposal tabled at the meeting also mentioned the possibility of utilizing lasers, or even a combination of systems.[53]

The second project was the design of inspection sensors and the development of techniques to determine satellite characteristics and functions. Deciding which characteristics were critical to the identification of a satellite or spacecraft, the COSINE reports proposed that the examination of techniques focus on technologies that would complement, rather than duplicate, the capabilities of other satellite-data-gathering platforms, both in space and on the ground.

The third, and perhaps most challenging, project dealt with non-nuclear kill mechanisms and negation techniques. At the time, there were still relatively few man-made objects in orbit, and the ramifications of destroying a satellite by breaking it into pieces and thus creating a dangerous cloud of debris were not yet fully appreciated. A number of kinetic methods were suggested, such as explosive warheads or shaped charges, as if the process were similar to that of engaging and demolishing an enemy tank. Although negation techniques were mentioned, they were discussed in much less detail. They included upsetting the position, spin, or attitude of the target, disrupting its orbit, or even corrupting its heat balance (almost all of which would result in the destruction of the satellite). Essentially, the proposal tabled at the meeting seemed to infer that, if the intent were to render a target useless, designing a cheap and simple self-destructing vehicle made more sense than creating an expensive, complicated craft that could interfere with the target but not destroy it. Likewise, nothing in the proposal mentioned capturing the satellite or taking control of it; hence, the tricky business of immobilizing but not permanently damaging a craft was not considered.

The Space Defence Program was divided into three projects so that, should one of them fail, the other two would not be jeopardized. Additionally, it was envisaged that each project could be executed in three phases. The first of these would generate detailed surveys and technical studies in the concepts of ballistic and orbital interception and co-orbital rendezvous. Because the United States was conducting similar research, it was recommended that members of the RCAF, the DRB, and Canadian industry gain wide exposure to a variety of American programs. The second phase would prepare preliminary hardware designs based on the studies undertaken during phase one and would build and test models on the ground. These would be assessed for potential development. The last phase would focus on fabrication and space testing of advanced models, after which operational equipment would be built and deployed.[54]

After meeting with the DRB on 14 May, the RCAF team met with two organizations that were most likely to become involved in the program: the DRTE (on 29 May) and CARDE (on 3-4 June). During these meetings, the RCAF proposal was introduced, and a detailed discussion period ensued. Follow-up visits were made to the DRTE and CARDE on 23 and 29 June, respectively, after both had had an opportunity to review the details of the program.[55] The objective of the meetings was to gain a better understanding of the resources and staffing costs required to implement the proposal, and to determine what was already available to undertake the studies that constituted the first phase of each project. Finally, a plan of organization was investigated, and a recommendation as to where the work should be done was made.[56]

Reactions by the DRB and its subcomponents to the RCAF Space Defence Program were less promising than perhaps expected. All three establishments – the DRB, the DRTE, and CARDE – expressed the general opinion that the program would be very expensive in terms of money and staffing. Though no detailed estimates were made at the time, a rough guess placed the program cost in the millions and the requirement for highly trained scientists and engineers in the dozens. Even this was an underestimation: similar high technology programs normally involved hundreds of specialized staff and support. Furthermore, general doubt was expressed as to whether the needed technical capabilities were available in Canadian government organizations and industry. By 1964 CARDE no longer felt that it had a strong capability in guidance systems, which had become painfully evident in its troubles with the early Black Brant III and IV trials. Likewise, the DRTE was uncertain whether it could provide competence to the Space Defence Program that would be similar to its ability in ionospheric research, and archival records imply that the DRTE had no desire to undermine its already substantial commitment to the topside sounder satellite program then under way.[57]

Concerns were also raised about the dependence of the Space Defence Program on the US defence space program and policy. If, rather than creating complete military interceptors or space vehicles for Canadian use, the program simply produced components and subsystems for American spacecraft, would the Americans actually employ them on a regular enough basis to justify the high cost of the program? Some critics asserted that the Space Defence Program did not contribute enough to Canadian scientific prestige, because its objective was to build subcomponents rather than whole systems. The RCAF team countered this argument with the observation that scientific and technological prestige was not a required by-product for a military program, and though the point was telling, the air force proposal won a Pyrrhic victory at best.[58]

The general lack of enthusiasm regarding the program proposal influenced the response to the initial study phase. The DRB was reluctant to commit high-calibre staff to the study workgroups without specific instructions from an authority such as its own chief or the chiefs of staff committee. It also suggested that the initial studies should not be restricted to the Space Defence Program but should be broadened to include other possible military programs. Again, this might simply have constituted a DRB attempt to make the program fit its own agenda, but nonetheless, it would commit effort only if directed from above.[59]

In a report that Flavin, Uhthoff, and Crysdale wrote after the various meetings, they noted that any attempt to resolve the differences between the RCAF and the DRB at lower levels would be futile and that the larger question of defining the Canadian military space program needed to be addressed at higher levels. Their report recommended that a steering committee of senior DRB and RCAF personnel should be formed to provide executive-level decisions and guidance regarding the future of Canadian military space operations. The committee would investigate relevant facets of American programs; conduct surveys of the state of the art; evaluate various program possibilities in terms of expenditure, requirement for available technical capability, and their impact on other Canadian organizations and industry; contribute to mutual defence agreements; and ensure compatibility of all defence programs with overall Canadian foreign policy and objectives. As no clear direction had been achieved during the meetings, Flavin, Uhthoff, and Crysdale obviously hoped that an oversight committee would be put in place so that, if a definitive space policy were issued, DND would be prepared to enact it. The recommendation was echoed by a senior air force officer, who commented that "the trepidation of recommending the formation of any committee at this point in time is overcome by the frustration of getting nothing done in the way of RCAF space activity in the last four years."[60] After so much effort by so many dedicated people, personalities were now threatening to unravel all that had been accomplished. Some officers, at least, could see the writing on the wall.

On 20 July, the RCAF-DRB working groups that had met to discuss all of these issues completed their reports and recommendations and sent them up the chain of command. The recommendations and apparent lack of decision provoked RCAF criticism. One officer noted with disgust, "If we are to opt out of future military technology let it be done by conscious decision rather than by default."[61] These and similar comments were duly collected and passed to Air Vice-Marshal Victor S.J. Millard, for forwarding to the chief of the air staff.

However, by this time, the integration of army, navy, and air force as the Canadian Armed Forces was already well under way, and the newly appointed air marshal C.L. Annis received the working group's final report instead. He sent both the report and the recommendations back down the chain of command, noting that, in view of the latest changes in DND organization and Ottawa's recent statements regarding the role of the Canadian forces, as outlined in its *White Paper on Defence,* the recommendations were misapplied and the past funding ($2.2 million) no longer assured. Although the 1964 white paper had devoted a small section to the issue of missile defence, its discussion of military space consisted of a single sentence congratulating the DRB for its launch of Alouette.

Though they were obviously germane to Canadian force development, generation, and protection, critical emerging space technologies such as navigation, communications, imagery, and meteorology were left out of the white paper. Its objective was the maintenance of adequate force to deter potential aggressors, a force that apparently did not include space-based military assets. At the time, Minister of National Defence Paul Hellyer was completely focused on the integration of the forces. Other large-scale projects, such as a military space program, were generally shunned by the senior leadership, which was already overwhelmed with the reorganization of defence writ large in Canada.[62]

Thus, C.L. Annis felt that an *ab initio* submission would be absolutely necessary if authority were to be gained to embark on a large-scale DND-sponsored space program. Therefore, on 30 September 1964, he directed that the Development and Associated Research Policy Group (DARPG), created specifically to address the issue, should prepare a new submission for consideration.[63]

Showing some sensitivity to concerns expressed earlier by senior RCAF personnel, Annis also wrote to Millard of his desire to bring more attention to the issue. In a memo written at the end of September 1964, he stated, "I am willing, even anxious, to see a fresh consideration of a DND space program," and he added that, in his opinion, the new military organization's decision-making machinery would facilitate future direction rather than hinder it.[64]

By the end of December, the new RCAF recommendation was ready for review by Annis, but due to changes in the organization of CFHQ, the final report never reached him and was returned to the DRB-RCAF working group. This time, the officer who replaced Millard stated that he personally would not be responsible for defence space programs in his new organization, and he suggested to the space defence advocates that they find a more appropriate channel to Annis. The resubmission was then passed through several deputies until it reached Annis, a hallmark of the ever-increasing bureaucracy at the integrating central defence headquarters. In the end, Annis supported only those recommendations that clearly defined an actual space program, not those associated with approving the three technical projects laid out in the 1965-66 defence estimates. After nearly seven months of effort, Canada's military space program had not taken a single official step forward.

In January 1965, the revised recommendations were submitted up the chain of command and finally reached the DARPG. Yet they were still considered inadequate, and the DRB's leading space development organizations, the DRTE and CARDE, embarked upon their own detailed national-level assessment of Canada's space activities to date. This separate effort ultimately led to the production of the previously discussed Chapman Report. The revised DRB-RCAF report and recommendations, along with other recommendations and interviews from various parties involved in space activity, were finally published as a report known simply as DARPG Paper 12/65. Some at CFHQ thought the report too little too late, but the context in which it was produced is worth additional consideration.

DARPG Paper 12/65

The Development and Associated Research Policy Group distributed its initial study on the Canadian Forces Space Defence Program in May 1965, providing a frank assessment of the challenges and triumphs to date.[65] It summarized Canada's military space activities and highlighted the generally disorganized defence approach to national space development, focusing in particular on the increased acrimony between the DRB and the RCAF since 1959.

It stated that, since the earliest days of the Cold War, the DRB had been the primary advocate for the development of a Canadian presence in space. It had collected and organized the various DND agencies that dealt with missiles and space and was largely responsible for early launcher development in Canada as well as detailed ICBM research and analysis. In 1957 it had initiated discussions with NASA on the possibility of launching a satellite, and by 1959 it was already at centre stage of Canada's infant space program and was devoting most of its energy and resources to its two primary projects – Black Brant and Alouette-ISIS.

The DARPG report pointed out that, until 1965, the three armed services had had limited involvement in the generation of space systems. Though this is true, it should not be taken to indicate a lack of interest in the military applications of space. As already demonstrated, the RCAF had a long history of participation in specific defence space programs. Furthermore, under the revised National Defence Act of 1947, the army, navy, and air force had been directed to turn over their research organizations to the DRB and thus had little direct involvement in the formulation of defence research policy throughout the 1950s and 1960s.

The military did participate in operational matters such as space tracking at RCAF Station Cold Lake and the Prince Albert Radar Laboratory but was otherwise largely relegated to the development of an air-transportable satellite communications terminal and a series of analytical studies on space rendezvous and surveillance. It also supplied the DRB with the majority of its test personnel, logistics support, and equipment such as specially modified jets for tracking incoming missile warheads.

In essence, the DRB was free to set its own research agenda and priorities, and as the DARPG report correctly pointed out, strong differences in opinion regarding the focus of Canada's space program had surfaced by 1965. Whereas the RCAF naturally sought military applications from its space program investment, the DRB, acting as a de facto Canadian space agency for nearly a decade, felt that the emphasis should remain fundamentally scientific. The DARPG report leaves little doubt that some senior DRB officials were upset by the entrance of an RCAF interest into the defence space agenda.[66] It noted in particular,

> Over the years, the various [Canadian] space projects have been approved in a piecemeal fashion and this has resulted in the creation of rather parochial thinking on the part of the various groups of experts. Thus in recent discussions, the DRB position on a military development program was conceived in terms of the anticipated next generation of research satellites to study the magnetosphere

FIGURE 40 A Canadian army signaller sets up a tactical satellite communications terminal pod during a Cold War training exercise. Such a system was made possible thanks to the country's growing focus on satellite communications development. National Defence Imagery Library, CF68-112-2.

> hundreds of thousands of miles from the Earth, as an extension of the ionosphere work at a few hundred miles; while the military continued to press for an extension of previous studies into research and development on military space systems.[67]

Had the spectre of the late 1940s RCAF-DRB debates risen again? Surely, the DRB's reluctance to incorporate RCAF space priorities was in part payback for earlier attempts by the RCAF to curb DRB efforts when it had greater control over defence research. Perhaps the DRB simply wished to protect the Alouette-ISIS project, knowing that resources were scarce and could not maintain more than one program. Whatever the case, one cannot fail to note the irony that the military's primary scientific research organization appeared to have little or no interest in making defence research its main objective. Undoubtedly, Black Brant and Alouette-ISIS furthered national interests, but the DRB was created primarily to support DND's preparation for modern war. This point was also highlighted in the report, which stated that "having regard for the basic responsibilities of this department [DND], there seems to be a strong case for a well-conceived research and development program oriented toward military satellites and space vehicles, rather than devoting the major part of our effort to a cooperative program with the non-military American space agency. This is not to degrade the value of the latter, but rather to bring into question whether it should be allowed to override the military need when funded out of the defence budget."[68]

Finally, the report highlighted the critical fact that, without an officially ratified civilian- or defence-oriented government space policy, further discussions on the direction of space research and development in Canada were academic. What was needed more than anything, DARPG advised, was guidance from the top. Interestingly, prior to the May 1965 release of the DARPG report, an April 1965 memorandum issued by J.C. Arnell, science adviser, made essentially the same point. As the senior non-DRB defence official involved in formulating Canada's space program, he wrote, "On reflection, I consider that until the basic policy decision is made, any detailed proposals for a program only tend to divert attention away from the basic problem."[69]

Perhaps the DARPG report's most important contribution was to identify the need for senior-level ministerial policy and direction on Canada's role in space. It certainly wished Canada's defence space program to be driven by military concerns but did not answer the question of who should assume responsibility for Canada's civilian and scientific space priorities if the DRB no longer took this role. As previously stated, the DRB was acting as the de facto national space agency. However, events during the next year would begin to change that.

Uncertain Future?

The DARPG study findings constituted a serious blow to the organization and direction of Canadian military space activities and caused DND's space advocates to re-evaluate their goals. Missile and space defence programs were becoming more complex, and the military was increasingly challenged to maintain saliency in such activities, as its resources and capabilities continued to shrink. Furthermore, the piecemeal approach to military space planning and operations simply highlighted CFHQ disorganization in such efforts. If this were not quickly addressed, the RCAF, and perhaps DND itself, was in danger of losing its space capabilities altogether.

At the end of September 1965, the science director to the chief of technical services called Lieutenant Colonel D.B.D. Warner, acting director of advanced concepts and systems, to his office "to discuss a number of items relating to Canadian military space activities, or rather the lack of them."[70] The conversation highlighted the fact that, within the massive integration and unification of the Canadian forces, the military space agenda was unravelling and suffering false start after false start. To instill some direction, the science director asked Warner to prepare yet another discussion paper outlining ideas that would help define requirements on Canadian military space activities. A month later, after conducting a series of interviews and investigations with the three services and several related agencies, Warner responded with a report.

Submitted on 28 October 1965, the report demonstrated what many within DND already knew – all military space programs were in trouble. Interviews with maritime, land, and air force combat development staffs reconfirmed earlier conclusions that very little thought was being devoted to future military space requirements. There were many reasons for this, not the least of which was the disruption caused by unification as organizations and their responsibilities regularly changed. An area that may have been assigned to an office one week was suddenly another's responsibility the next. Furthermore, those who were given space-related tasks were often ignorant of the possible uses of space systems and therefore unaware of the requirements needed to exploit them.

An example of this appears in a 1963 memorandum prepared for the chief of the naval staff, which stated, "while the RCN has no very pressing direct interest in this field [space] at present, one may expect that in due course it will yield developments of importance to the navy ... It has been recommended that no direct RCN participation be undertaken for the present, chiefly because suitably qualified personnel are in too short supply to be spared at this time when the naval interest is unspecific."[71]

Given the plethora of problems facing combat developers at CFHQ, space activities were accorded little priority if not ignored outright. Finally, Warner noted that "space activities as a whole imply very large expenditures and this stigma tends to be attached indiscriminately to any space proposal."[72]

It was increasingly apparent that no one in the military was anxious to become a champion for outer space. Even the RCAF, the primary advocate and proponent of military space programs for many years, was beginning to withdraw its usually high level of support. None of the three services' combat development staffs had the ability to instigate broad or detailed requirements for space programs; neither did they propose to establish any on their own initiative. "The general attitude toward military space activities must be summarized for the time being," Warner wrote, "as disinterested."[73] He concluded, "In this environment it is unlikely that much support would be given to space proposals of a long-range nature, and particularly not to proposals that suggest new roles for the Canadian Forces. It is much more likely that proposals directed toward the enhancement of existing and anticipated roles will survive. This is a case of first things first and suggestions probably should be confined to this area if there is to be much hope at this time of meaningful VCDS [vice chief of defence staff] support."[74]

As the Canadian forces marched toward unification, their missile and space programs, like so many other technologically advanced ventures, suffered accordingly. Little political direction existed to make space a defence priority, and the generals seemed unwilling to stake their own interests on high-risk programs that might never come to fruition. Also, Canada's scientific and technical capabilities could not keep up with the increasing complexity of modern weapon systems and space platforms, which meant that existing programs were degraded and future proposals progressively more unlikely. These factors, combined with the lack of space education and the ever-increasing budgetary constraints, posed serious challenges to the survival of Canada's military space programs.

6

The Demise of Canada's Cold War Space Programs

In 1965 Canada's military space programs were in jeopardy. Obstacles to the formalization of the military space agenda persisted, and plans for steering existing scientific and technological capability in the direction of military interests had yet to materialize. Similarly, efforts to transfer those technological requirements from defence to civilian industry largely failed, challenging the assertion that military establishments played an important role in shaping technological and economic change by linking national defence with national welfare.[1]

Widening every year, the Canadian "technological capability gap" simply divided political interests within the defence community as the Defence Research Board scrambled to protect its much-loved Alouette-ISIS program at the expense of military efforts. In the agitated climate of a dynamic shift in Canadian foreign policy and the reorganization of the defence and scientific communities, support for missile and defence space endeavours evaporated. One by one, the armed services withdrew their investments from these programs, which were seen as costly and increasingly technologically impossible to develop and sustain, in order to save others. By the end of the 1960s, the missile and space programs were all but terminated as Canada's military assets were transferred to civilian government sectors and much-needed resources and funding were allocated elsewhere.

Altering Course

Canada's defence policy and establishment underwent a radical transformation during the mid-1960s. During the Cuban Missile Crisis, John Diefenbaker's Conservative government had demonstrated a considerable inability to exercise positive control over the country's armed forces, realizing almost too late that the military was, as historian Douglas Bland notes, "sharply focused on alliance duties and largely responsive to allied commanders."[2] Soon afterward, it was suggested that Ottawa had no national plans for war fighting, maintained only a weak indigenous intelligence capability, and had no reliable structure for commanding and controlling the armed forces as a whole. Subsequently, the Conservative government was again embarrassed by its handling of the country's seemingly schizophrenic nuclear weapons policy, and further political fumbling revealed Diefenbaker's surprisingly feeble grasp of national defence policy. In the end, this series of events broke Cabinet solidarity and ultimately destroyed any chances of a Conservative Party re-election in 1963.[3]

The Liberal Party's return to power brought considerable changes to national security priorities, which affected the country's defence apparatus at every level. The new prime minister, Lester "Mike" Pearson, had no desire to repeat the mistakes of his predecessor and immediately took action to ensure that similar problems did not plague his own Cabinet administration. He appointed his tough-minded defence critic Paul Hellyer as defence minister and instructed him to establish firm control over the Department of National Defence while at the same time reducing defence expenditures.[4] Hellyer was more than equal to the task, and his tireless efforts to integrate the Canadian forces under a single headquarters and to reshape DND during the next four years made him one of the most controversial defence ministers in Canadian history.

The Pearson government had made many expensive promises during its election campaign, including the establishment of health care and old age pensions for all Canadians. To generate funding for these ventures, Pearson directed his finance minister, Walter Gordon, to cut other areas of domestic spending wherever possible. Naturally, the prevailing sense of international stability after the Cuban Missile Crisis and the signing of the Nuclear Test Ban Treaty in 1963 made the defence budget an obvious target. Gordon consulted with Paul Hellyer on the matter, and the two agreed to freeze defence spending at $1.5 billion for the next three fiscal years.[5] This resulted in the immediate cancellation of plans to procure frigates for the Royal Canadian Navy and additional CF-101 Voodoo and CF-104 Starfighter aircraft for the RCAF. The focus of the department was now the department itself, and Hellyer made it clear that all new programs and acquisitions were on hold until a comprehensive review of defence policy and administration was completed.

The review itself was carried out in many parallel forms, from public hearings to confidential assessments.[6] Within DND, Hellyer had asked R.J. Sutherland, chief of operational research at the DRB, to undertake a confidential and extensive study of Canadian defence policy. When this was completed, Cabinet indicated that the government's new policy would emphasize Canadian-oriented priorities and the likelihood that future armed conflicts would require a flexible and conventional response rather than a massive retaliation against large-scale Soviet nuclear attacks. Anticipating that tensions would persist in Europe, albeit at lower levels, Ottawa rightly predicted that Canadian security could be best achieved via a level of involvement in NORAD and NATO, a pragmatic negotiation with the Communist bloc, and a contribution to the containment of Soviet influence worldwide.

That Canada espoused this logic should come as no surprise. Paul Hellyer subscribed to the new NATO concept of a flexible response to Soviet military threats, as it suggested the need for change. He also followed the lead of his American counterpart Robert McNamara in trying to discipline the senior leadership of the military and to exert stricter control over its spending. Finally, flexible response also provided a strategic rationale for Canada to maintain a conventional capability while accepting the nuclear weapons that the Pearson government had agreed to add to its arsenal.

However, developing a flexible response did not necessarily require large investments in missile and space programs. In late 1964 and early 1965, Ottawa further distanced itself from investing in continental defence when it made no provision for the acquisition of equipment designed exclusively for nuclear arms and did not solidly commit itself to modernizing strategic defence.[7] Instead, it leaned toward an increasing role in international peacekeeping and the provision of military assistance to developing nations in Africa and Asia.[8] At home, it was satisfied to retain Canada's aging fleet of CF-101s and newer CF-104s, while making increasingly marginal contributions to the development of the North American security infrastructure. Internally, the government already accepted that defending North America from nuclear attack or invasion would fall largely on the shoulders of its American ally.

These changes did not go unnoticed south of the border. Although it did not disapprove of Canada's growing involvement in international peacekeeping, the United States hoped that this would not supersede Canadian contributions to North American strategic defence. Walt Butterworth, the American ambassador in Ottawa at the time, complained that Canada's defence policy showed disturbing signs of neutralism, and Washington reminded Ottawa at every opportunity that its "revised defence policy does not impair Canadian commitments to [the] joint defence of North America and NATO."[9] Although Washington realized that it would ultimately have to reassess its policy concerning Canada, it strongly encouraged Ottawa to retain the present levels of Canadian forces deployed with NORAD and in Europe with NATO.

Regardless, little could be done to halt the atrophy in Canada's strategic defence initiatives. The country simply could not afford to sustain the technological or financial resources needed to field a large modernized nuclear-capable force. Consequently, it was obliged to sacrifice major roles in strategic defence, choosing to favour a reasonably high degree of conventional punch with the savings that accrued from the unification of the three services. Though politically judicious, this decision did not bode well for those military projects aimed at building a defence space capability in Canada.

Losing Momentum

The publication of the Chapman Report left the DND Space Defence Program proposal without real representation in Cabinet when the time arrived for a government decision on the focus of Canada's space efforts. Internally, DND had assessed the status of its missile and space activities. Although senior RCAF leadership had voiced its concerns regarding the dangerous Soviet threat via space, RCAF influence in defining Canada's defence priorities had seriously waned during Paul Hellyer's tenure.[10] Funding was also slashed, augmenting growing apprehension among RCAF planners that several major missile- and space-related programs and projects were slipping and that sustaining and expanding them would receive inadequate attention throughout the rest of the decade. If programs were to be cut, the RCAF needed a candid top-level assessment of its activities to ensure that the projects absolutely vital to the strategic defence of Canada avoided the political chopping block.

The RCAF had conducted a series of performance reviews beginning in early 1964, the same year that the first of its Space Indoctrination Program (SIP) officers returned home from their assignments in the United States. Their impressive reports on American missile and space achievements highlighted the continuously widening gap between Canadian and American technological capabilities, but the internal reviews truly demonstrated the growing frustrations and flaring tempers in the RCAF.

First to be scrutinized was DND's diminishing contribution to bilateral ballistic missile defence research and development. In March 1964, Canada had formally withdrawn from Operation LOOKOUT (and in a larger sense from Project DEFENDER as a whole) despite recommendations from the United States and the DRB that it remain committed to this research. "Since we have held ourselves to be a service which is anxious to contribute to western and North American defence," wrote one senior RCAF officer, "it is difficult to understand why we withdrew from an area of such vital interest as [Project] DEFENDER simply on the grounds of a manpower and economy problem which was not large."[11] Commenting on whether the withdrawal reflected an intent to redirect the RCAF's focus from strategic defence to more compact streamlined support for international policing actions and conventional warfare, the officer wrote, "it would follow that we withdraw from any effort related to AICBM [anti-ICBMs] or space."[12]

Canada's shrinking contribution to SPADATS was the object of similar scorn. Since entering the business of space surveillance in 1961 with a single Baker-Nunn camera and the Satellite Identification Tracking Unit at Cold Lake, the Canadian military had not augmented its SPADATS capability. Though an additional computer had come on-line at the DRTE to provide limited assistance with data processing, the processing demands were rapidly exceeding RCAF capacity. Equally distressing, the Baker-Nunn camera itself was becoming obsolete as more modern electro-optical systems were produced. In addition, physically processing and transferring data to punch cards for computation was slow and antiquated; newer electro-optical devices could transfer electronic images directly into digital storage. Senior military leadership was aware of the obsolescence of the Baker-Nunn and its data-processing techniques. A 1964 review of the program stated that "the Baker-Nunn Look Angle Data project is effectively dead and should remain so."[13] Given such comments, one wonders what tangible contribution Canada did make to the surveillance of space after 1965.

Canada's SPADATS project in particular highlighted the piecemeal fashion in which DND had participated in the militarization and weaponization of space throughout the 1960s. When the Air Defence Command and technical staffs at CFHQ attempted to improve SPADATS by modifying and relocating the Baker-Nunn camera in 1965, this forced a re-examination of the DND requirement for involvement in space detection and tracking. The examination revealed that no clearly stated requirement actually existed, and more disconcertingly, that it had never been sanctioned by Minister of National Defence Paul Hellyer.[14] As a result, not only were the improvements halted but the entire contribution was brought into question and challenged. It was temporarily saved only when the combined combat development staffs recommended that it be preserved and the improvements made. Nonetheless, when their recommendation went before the Defence Council, it again stalled, until eventually technology and events rendered the matter obsolete.[15] As predicted in 1964, SPADATS was "effectively though not formally dead."[16]

Other military space programs declined. Though the second rotation of SIP officers was dispatched to the United States starting in 1964, those who returned to Canada after several years in the continuously expanding American space program were considerably disappointed by what they found at home. As one SIP officer noted, "when three of us came back to Canada in 1965, we were gathered together and briefed on thoughts that were floating around headquarters on how DND might be involved in future space activities. We were all so dismayed by the lack of knowledge and understanding, that all three of us clearly indicated that we wanted no future involvement in such DND plans."[17] To those who had served elsewhere, the deterioration of advanced missile and space capability in the Canadian military was painfully obvious.

Not all space-related projects within DND experienced setbacks. By the mid-1960s, the newly integrated Canadian Armed Forces were serving around the globe, in Europe, the Middle East, Africa, and Asia. Keeping a link between such widely dispersed forces could be achieved only with robust and sustainable wireless communications such as those provided by satellite. Because DND already had experience in satellite communications, less effort and new funding were required to develop this much-needed capability for the forces. As a result, satellite communications slowly progressed, whereas other space projects that demanded more resources were stunted.

Even so, Canada's nascent satellite communications capability struggled because of missed opportunity and the inability to diffuse military technology to civilian industry. The air-transportable satellite communications ground terminal, identified by the CFHQ combat development staff as necessary for the forces, never evolved in Canadian industry. Meanwhile, by 1965 the United States had contracted to a company named Radiation Incorporated for similar terminals, and eventually DND accepted them for Canadian use. "If terminals are to be acquired by Canada," stated a DND report in late 1965, "they will no doubt be purchased from the US."[18] But buying from the United States meant waiting until all American orders were filled. The Canadian forces would not acquire their terminals until some time in mid-1968, nearly three and a half years after the product came off the American assembly line.

Furthermore, DND assumed that no terminals would be purchased until an agreement or memorandum of understanding existed between Ottawa and Washington with respect to Canadian use of the US military satellite communications system. Although Canada had launched two experimental communications satellites by 1965, it was still years away from purchasing and deploying its own space assets. Except for the Soviet Union, the United States was the only country with orbiting platforms that could support satellite ground terminal development. Therefore, employing American satellites made sense, but the political and legal frameworks needed to be in place before technical development could ensue. Additionally, in late 1967, a NATO Tactical Satellite Communications Steering Committee was finally established, followed by a technical testing program in 1968. Canada participated in both these projects, achieving considerable success.[19]

Canada's employment of space-derived meteorology products likewise depended greatly on access to early American satellites such as TIROS and NIMBUS. TIROS satellites provided daytime visible-light pictures of cloud cover and weather systems, whereas the NIMBUS satellite added nighttime infrared pictures of these and of the earth's heat distribution, which was especially important to weather forecasting in the 1960s. A small group at the NRC, working under Charles Taggart, who himself was seconded to research from the Department of Transport, designed experimental terminals for relaying space-derived weather information using a process known as automatic picture transmission (APT). This technology proved to have considerable military application. For example, the US Armed Forces had about fifty APT units in operation around the world during the 1960s, including one in Vietnam.[20]

Taggart's prototypes were very successful, and for once, Canadian industry was induced to step in and manufacture an advanced product. Working under contract, Canadian Aviation Electronics built two stations, one at Halifax, Nova Scotia, and one at Milton, Ontario, to provide improved weather forecasting capabilities to both military and civilian agencies. These worked well, and the company submitted further proposals to install APT terminals on HMCS *Bonaventure* and other Canadian warships. With little or no action on its part, the military had gained considerable capacity in this field, and at a small cost, provided that the agreements with the United States could be kept in good standing.

Despite these achievements, however, the Canadian military space program continued to lose momentum. Though the RCAF did maintain a few smaller projects primarily focused on the exploitation of space-derived data, it was unable to realize any large-scale projects such as spacecraft or satellite research and development. Changes at CFHQ in 1965 and 1966 resulted in some reorganization of operational research, including the oversight of military space activities, but this did not improve the situation. The reshaped Directorate of Strategic Weapons and Air Defence was given the mandate of "maintain[ing] cognizance of developments in the field of strategic weapons including the military applications of space" but was not empowered to implement programs leading to the deployment of military space assets.[21] A 1966 reference paper on the Canadian forces role in the upper atmosphere and space reaffirmed a general policy commitment to many of the traditional objectives of the RCAF Space Defence Program, but by this time any effort spent sustaining that intangible proposal was probably recognized as pure folly.

Endgame: The Canadian Military Space Group

Despite the fact that John Chapman, who led the Science Secretariat task force that reviewed the national space program in 1966, was himself a defence scientist, his mandate from the secretariat was to focus on civilian issues only and not to investigate military space programs. Although the two fields were not mutually exclusive, given the dual-use nature of Canadian space organizations and assets, space advocates within DND realized by the spring of 1967 that the Chapman Report would not recommend much support for the development of a military space program, especially if this might compromise Chapman's own legacy in space communications research.

Though not completely unexpected, given that the recent upheaval in science, technology, defence, and government threatened all space-related agendas, the Chapman Report's conclusions and recommendations still caught the senior defence community somewhat unprepared to address the issue at higher levels. Since the beginning of the space era, the DRB had acted as a primary agent for Canada's defence space program in Cabinet, but now it seemed that Chapman's impending report would eclipse all other views, including that of the military. The situation sent DND racing to formulate a position for its own space activities, outside of pure scientific research.

Although the Directorate of Strategic Weapons and Air Defence was charged with monitoring developments in space technology, it was not tasked with generating defence space policy or plans for the department. That responsibility fell to the office of the chief of technical services (CTS), which was previously involved with the promulgation of the RCAF Space Defence Program and now merged with the department's directorate of science.

On 12 May 1967, Vernon Smith, special assistant to the CTS at CFHQ, submitted a position paper titled "A Canadian Military Space Program – The Need for Action" to his fellow special assistant, R.F. Wilkinson, for review.[22] Smith had read the Chapman Report prior to its submission to the Science Council of Canada and was gravely worried about its possible implications for military space programs. His own paper was a candid statement, and it made no effort to hide his disappointment in the Chapman Report's conclusions or to downplay the precariousness of DND's position. Smith's analysis is worth quoting at length:

> The disappointment within CFHQ occasioned by the mere passing references to a Canadian military space program in the Special Study No. 1 [Chapman Report] ... was natural, for a number of competent vigorous officers have been engaged for a number of years on programs allied to space. Unfortunately, however, the absence of a cohesive policy to exploit space for Canadian military ends resulted in the submission which could be no more than a parade of past and present projects related to "space" as defined in its broadest sense. But this is history and there is little time to lose if we are to recover the lost ground. It is the purpose of this paper to propose an immediate modus operandi, which will allow us at best to take the initiative, or at worst to hold our own in the competition for space resources and funds within Canada.
>
> The urgency arises because the Science Council is expected to approve the Chapman Report at its meeting on 8 May, and shortly thereafter publish its findings. The recommendations will be directed to the Cabinet, and should they be

> approved, the next step will be the planning, preparation of legislation, estimating of costs, etc. necessary for the establishment of a "space agency" in perhaps two years time.[23] Hence planning may well start this summer or fall, and the absence of a DND position which the Minister can support could be serious. Not only would we have no slice of the money pie, but the agency will undoubtedly cast about for facilities and misemployed specialist officers within the DND whom the latter will be obliged to surrender to those who can better employ them. Thus we should make every effort to have a rational, even if incomplete, policy by the end of July so that the Minister may have a military position compatible with the overall national policy.[24]

On 19 May, concurring with Smith's report, Wilkinson forwarded it to the CTS, Lieutenant General L.G.C. Lilley. However, Lilley completed his review of the document only on 5 July, almost a month after the Record of Cabinet Decision regarding the Chapman Report had passed. Thus, DND missed any last-minute opportunity to influence Cabinet decisions arising from the Chapman Report and now found itself scrambling to insert itself into the new organizations that evolved from the report to implement Ottawa's new vision for the national space program.

On 21 July, DND representatives had their first opportunity to meet with the newly created government Ad Hoc Task Force on Satellites, when Colonel M.H.F. Webber provided a presentation to them outlining Defence Chief of Review Service issues. R.C. Langille from the DRTE then gave a brief account of DRB programs, and he and Webber stressed DND's interest in satellite communications projects. Whether this was simply a means of currying favour with the task force is unknown, but the records do suggest that DND was attempting to create linkages between defence and civilian space programs. Such initiatives constituted the beginning of a concerted effort to save the defence space program.

Lilley added his own voice to the issue on 9 August, when he forwarded Smith's paper to Lieutenant General H.L. Meuser, the assistant deputy minister-logistics, and requested that Meuser send someone from his office to an ad hoc group known as the Canadian Military Space Group (CMSG) that Lilley's office was putting together. The group's aim was to formulate an official DND space policy and to promote it within the government's Ad Hoc Task Force and its successors. In particular, it would update the recommendations put forward by Vernon Smith in his 12 May position paper and provide a swift reaction to questions affecting DND that were generated by planning and policy advisory groups such as Chapman's task force.[25]

Meanwhile, Lilley's office continued its efforts to build support. Wilkinson sent an urgent memorandum to the deputy chief of engineering in which he wrote that "the recent events surrounding Dr. Chapman's 'Special Study on Upper Atmosphere and Space Programs in Canada' and his Ad Hoc Task Force, both of which are familiar to you, convince me we must move ahead quickly with developing a DND position and providing a fast reaction to government planning bodies."[26] Similar notes of urgency were forwarded from Lilley's office to other offices within DND. Within the week, they had begun to select the men who would represent them on the newly created CMSG, which was expected to meet for the first time in early September.[27]

For the CTS, the main challenge to gaining a foothold in the new Canadian space program was time. DND had no overall strategy for space let alone a policy, as it had never had to think or express itself in these terms. Traditionally, Canada's military space program, albeit complex, was designed to knit into the American strategic defence of North America. Lilley was therefore under no illusion regarding the danger of the department's position. If he did not act quickly, DND might be shut out of space activities altogether. In acknowledging this, Lilley noted to Meuser that "events have overtaken us ... The government is under such intense pressure to reach decisions that [Chapman's] interim findings will be available [for implementation] by 1 September."[28] For example, Cabinet had already ratified a Canadian satellite communications initiative in the summer, and government planning for it had commenced. Since domestic policy in Canada took priority, it was likely that other attractive civilian space programs would be passed by Cabinet with similar expediency. Soon, available funding and resources for space would be fully committed, and Lilley feared that DND, caught unprepared, might receive none of it.

The Canadian Military Space Group met for the first time on 15 September 1967, in the CTS conference room at CFHQ.[29] It was an impressive assembly of some of the department's most experienced experts, but all in attendance must have privately wondered if their meeting had come too late. Nevertheless, they were committed to the task at hand, and many of them would see the process through to its end years later.

Seven men sat around the conference table. Lilley and his special assistants (Vernon Smith and R.F. Wilkinson) represented the office of the CTS. Others included Lieutenant Colonel L.H. Wylie, whose organization was already active in managing the technical and tactical testing of the LES-5 satellite system, and Group Captain P.F. Peter, who was responsible for military aerospace activities. F.S.B. Thompson, director of communications requirements, represented the assistant deputy minister-logistics, and Roy Doohoo, acting superintendent of the communications laboratory at the DRTE, attended on behalf of the DRB. Colonel M.H.F. Webber, from the office of primary interest for communications-electronics activities at both the strategic and tactical level, was invited to attend but could not appear at this first meeting.[30]

Lilley gave a brief keynote address in which he stressed that DND must have a space policy compatible with that of the government, as expressed by the Science Council and associated planning groups. Lilley believed that DND's interests could be best served by cooperating with various other government departments and agencies on space projects, thus allowing the military to remain involved in activities that could serve a dual use. Finally, he recommended that the group assembled before him complete a study within six months, with a first interim report ready for review in approximately two months.[31] Time remained a key factor in determining success or failure.

A number of other issues were also discussed. Attendees noted that they could not undertake their own specialized tasks related to military space program development until overall DND requirements were established, and thus had to wait for a plan to be put in place. Though unable to wait for direction from the yet to be established Canadian Space Agency, the CMSG estimated that DND's aim was to develop a low-risk state-of-the-art program within the capabilities of Canadian industry. The meeting therefore stressed that, whatever action plan was pursued, a full awareness of government intentions regarding space was essential if an effective DND space policy were to be built.[32] Finally, terms of reference were tabled, as was a tentative agenda for the next meeting. With business done, the group dispersed for the moment.

After two months of gathering data and strengthening its position, at least on paper, the CMSG convened its fourth meeting on 2 November. William M. Searle, who had joined during the third meeting, was appointed its secretary. After completing opening business, each member of the group reported on his latest activities.[33]

Overall, the reports were not very promising. Whether or not Canada needed to parallel US space development, it could not duplicate the infrastructure being created in the United States, an observation made years before by the RCAF study groups. In any event, Cabinet seemed increasingly unlikely to support a large space organization.[34]

This latter observation came from General Lilley, who reported that, during a recent meeting with John Chapman, the latter had stated that the new space program he proposed was not progressing well. Chapman had added that, though Cabinet was likely to almost immediately table decisions based on his report, he felt that only a diluted version of his recommendations would pass and that many issues remained unresolved. For instance, funding was a huge problem: though planning for the Canadian Space Agency would probably begin at the end of the year, Chapman told Lilley that its funding remained to be determined.[35] Funding would also be needed to build, operate, and maintain a space infrastructure and industry in Canada, and nothing indicated when or even if it would ever be available on the scale that Chapman desired. Perhaps most important to the CMSG was Chapman's comment regarding Canadian military projects. He announced bluntly to Lilley that, as per his mandate, his report would not affect the military at all, but realistically, any military requirements, even communications, would probably remain secondary to the more apparent needs of domestic policy and the civilian economy.[36]

Though he had already thoroughly depressed the CMSG, Lilley relayed the details of a meeting he had had with Major General D.A.G. Waldock. The latter had wondered whether, given the shortage of money and staff, the CMSG was trying to press too far ahead of government on space issues. He recommended that, instead of trying to set out plans for a full-blown military space program, the CMSG should keep the technical team involved to determine DND requirements as well as how much funding would be available in two years' time for satellite projects and programs.

Given this development, Lilley suggested that the CMSG should immediately prepare a report describing the current situation. L.H. Wylie disagreed with the need for speed, arguing that if the CMSG was planning for the 1970s, there was plenty of time for discussion before submitting a final report to senior DND leadership. However, Colonel Webber countered Wylie's argument, concurring with D.A.G. Waldock's assessment and adding that he too felt it unwise for DND to try to get ahead of the proposed space agency program. Finally, after the debate concluded, Lilley tabled the motion that, given the uncertainty of the situation, the CMSG should make recommendations regarding its own future.[37] With that, the meeting ended, and the group dispersed once more.

Throughout the winter of 1967-68, the Canadian Military Space Group continued its efforts to prepare a comprehensive report for the spring. Further meetings were held on 16 November and 21 and 28 December. Activity centred on preparing the committee's report, which was the subject of a detailed review at a meeting on 18 January 1968. Group Captain Peter, now a veteran space advocate in the Canadian military, felt that the draft report still provided too much detail and thus missed the larger objective of defining a Canadian defence policy or program. Roy Doohoo voiced similar concerns in noting that the report focused not on how existing systems could be exploited but rather on convincing the government that DND required its own satellites for use.

Again, very similar arguments had already been made, when the DRB and the RCAF met in the mid-1960s. At that time, the DRB had found a possible way ahead through the Chapman Report, but the military, left without a champion in Cabinet, became increasingly frustrated at its own lack of progress. More importantly, by 1968, many senior military men who had forged the way ahead for Canada's military space program back in 1959 had either taken on other tasks or left the military altogether. There were not many left like Group Captain Peter, who had been involved with the RCAF Space Defence Program from the outset and who still believed that the Canadian military could greatly benefit from an indigenous program. Instead, although DND was disappointed with the status of its efforts, records suggest that, toward the late 1960s, it was increasingly realizing that attempting to establish a space policy was pointless, given the current upheaval in both the defence and scientific communities. Thus, any hope of building a space agenda for DND in the 1970s lay within the pages of the Canadian Military Space Group's 1968 report.[38]

The first report of the CMSG, prepared on 31 January 1968, was submitted through L.G.C. Lilley to the chief of the defence staff at the beginning of March. At just over thirty pages, it contained a considerable amount of detail but failed to emphasize what all the military space advocates had said in conference over the past years – that a defined defence space policy or vision was required. Although its introduction was clear in stating that the Chapman Report made "only passing reference to the CAF [Canadian Armed Forces] and conveyed little or no impression of their interest in space activities," it did not address this oversight in its executive summary.[39] Instead, it described its aim as twofold: first, it would convey the CMSG findings regarding the impact of Canadian defence requirements on proposals for Canadian national programs; second, it sought guidance on the CMSG's future activities. It did not deal with the need for an unequivocal position on military space, and the section on future DND space policy, buried in mid-discussion, stated weakly "that the time is not appropriate for the establishment of Canadian military policy concerning space activities relative to other DND commitments."[40] Essentially, if its intention was to convince the new chief of the defence staff to make a physical commitment to a military space program, the report failed in its opening paragraphs. In seeking to solidify the CMSG's terms of reference, the report suggested that one of its tasks should be to formulate a course of action that would satisfy the defence space program needs of the Canadian Armed Forces after 1972. After fourteen months of meetings, the CMSG was unable to devise a credible plan of action, a fact that must have produced some skepticism among the senior leadership as to how much it could expect to accomplish, given its size, mandate, and available resources.

How are we to explain the failings of the CMSG report? Although many of the officers who were most experienced in rocketry, missile, and space activities were no longer involved in these areas or had left military service, some remained. Yet neither they nor SIP exchange personnel were fully consulted prior to the drafting of such a critical report. Obviously, the CMSG worked under time constraints, and undoubtedly the organizational upheaval within DND had some impact, but placing the hopes of future DND military space activity on such a frail vessel seems irresponsible. In the end, one can only assume that perhaps the CMSG members themselves lacked the education or sophistication to "sell" military space projects to their political bosses, or perhaps they felt the effort was futile from the start.

Response to the report and its recommendations came quickly. Within a week of its submission, policy guidance was distributed from the new director of strategic force planning that noted, "At the present time there is no clear cut Canadian military requirement for the use of space technology" but that the Canadian Armed Forces should continue to monitor space developments as they occurred.[41] Likewise, though the ministerial-level defence policy review studies then under way did not touch upon satellite or space capabilities, the director decided that there was no need to modify those studies to reflect this omission. Rather, he deemed it sufficient to include a note on the requirement for secure communications in the command and control section of the studies.[42] Canadian military space ventures would be dependent on a government decision to embark on a national space program at some point in the future.[43]

The remaining CMSG meetings of 1968 centred on presentations that dealt with Canadian satellite communications and their potential military applications. The last meeting of the year, held on 5 September, focused on communication satellites and Canadian industry. Sadly, it was attended by only nine people, including the two invited presenters. Plainly, any enthusiasm for what appeared to be a hopeless case had vanished, especially after the group's efforts to draw attention to military space activities had resulted in so few responses both within and outside the department, and nothing in the way of a commitment or a champion. After the 5 September meeting, the CMSG would not reassemble for almost a year.

In June 1969, Wilkinson distributed a brief note to CMSG members, calling for a meeting at the end of July. This session attracted more people, but many of the original group had left CFHQ or DND altogether, resulting in new faces around the table. Vernon Smith, one of the original instigators of the CMSG, had transferred from CFHQ to the Telecommunications Liaison Office at the Defence Research Establishment (Ottawa) in 1969, but he returned at Wilkinson's request to chair the meeting. It consisted primarily of a lengthy presentation on Canada's non-military space program by David Florida, now working at the Communications Research Center. Otherwise, the minutes record no discussion of Canadian military space activity, leaving one to wonder why the meeting was held.[44]

Although the issue of terms of reference and the CMSG's future mandate had been previously discussed, nothing indicates that these matters were addressed in any detail at the July meeting; nor was a date set for the next one. Later, a 28 August 1969 departmental memorandum on the future of the CMSG signed by C.R. Iverson, special assistant to the CTS, did not mention terms of reference for the group but did declare that the CTS agreed that it should meet annually or at the call of the chairman, presumably either Wilkinson or Iverson himself.[45] There is no evidence within declassified files to suggest that the CMSG ever met again.

Military Imperative and Technological Challenges

As mentioned above, the exploitation of science and technology in preparing the defence of North America during the Cold War presented Canada with considerable challenges.[46] As early as 1947, the more technically oriented among Canada's senior military planners realized that the absence of sophisticated electronics, computation systems, and miniaturization would probably doom any Canadian effort to secure a degree of self-reliance in the development of advanced early warning systems and terrestrial and space-based military platforms.[47]

By the early 1950s, the Canadian military faced increasing challenges in keeping its force modernized while at the same time meeting the DRB and RCAF proposals for missile and space programs. The proliferation of electronics that revolutionized military access to space also threatened to compromise the effectiveness of these programs as the scope and diversity of electronics systems grew. The Canadian civilian electronics market, dominated by phone, radio, and television production, used small amounts of large vacuum tubes, but military performance specifications for electronics demanded far higher quality than this. Thus, the military was forced to employ its own defence laboratories to finance the costly development and production of needed items.

There was no avoiding the high demands for efficiency and reliability created by increasingly complex digitally based platforms such as missiles and spacecraft. For example, the vacuum tube in a radio or television might run for approximately 87,660 hours before it failed. Thus, on average, that component would fail once every ten years. This might have sounded impressive to a potential military client, but his computer, missile, or space system might require ten thousand components. If such were the case, and it often was, his system was likely to fail every eight or nine hours. This was not a great prospect for advanced missile systems, and it was absolutely unacceptable for a spacecraft that, once launched, could not be retrieved for repair.

The Canadian military's move from the employment of vacuum tubes to, first, germanium and later silicon transistors, which was brought on by DRB research, was continually hampered by those who feared change. Apprehension regarding unproven technology combined with reluctance to switch to electronics, which were sure to fail in a nuclear war (as nuclear blasts would knock out transistors but not vacuum tubes), ensured that military imperatives were not quickly met.

The need to deal with the rapidly increasing complexity of electronics used in weapons systems, fuelled by a desire for self-reliance, led the military to focus on promoting a domestic capacity to miniaturize. Yet, for a number of reasons, advances made at the DRTE transistor section did not carry over to civilian industry, forcing DND to provide for itself. DND efforts to subsidize Canadian industry were criticized, even from within the department. One deputy minister of defence questioned the legitimacy of expending public funds to foster industrial development, and as defence budgets vanished after the Korean War, DND attempts to support the domestic production of electronic components diminished until the enterprise simply became too expensive.

For example, without civilian demand, transistor production remained very costly. A transistor could cost eight times as much as a vacuum tube, offering little incentive for civilian electronics markets to incorporate it into their products. Also, as historian John Vardalas points out, there was no technical imperative for the civilian market to switch to solid-state miniaturization at the time. Finally, the move to transistor technology would undermine the large capital investments already existing in vacuum tube manufacture, a fact that simply strengthened social and economic resistance to the diffusion of this technology and further impeded the process.[48]

The AVRO CF-105 Arrow interceptor jet experience certainly underscored this problem. Although the Department of Defence Production had voiced its concerns over the lack of investment in miniaturization and its imperative to military self-reliance as early as 1951, no serious effort or commitment was made to induce a domestic commercial market. As a result, the Arrow depended largely on American-built miniaturized subcomponents, and ultimately the project itself became unsustainable when other American priorities and capabilities superseded it.

Lacking an organized capacity to manufacture electronic components, DND became increasingly concerned regarding whether Canada could fulfill its responsibilities within the NATO alliance to design, test, and construct transistor-based equipment. Though one could argue that this equipment would have made little difference in the event of a nuclear war, building and deploying advanced military platforms before war came was still considered necessary. Yet the lack of such an indigenous capability would eventually affect Canada's entire defence industry.[49] And though politicians were seldom interested in the details of defence, this area had ramifications for the civilian economy as well. Failure on the defence technology front meant two things. First, Canada might jeopardize its alliances by failing to meet its defence commitments, which would in turn affect trade and commerce. Second, by not supporting the diffusion of technology from defence to the civilian economy, the government was also affecting overall national welfare.

If the RCAF could not ameliorate the electronic components issue, Canadian military missiles and spacecraft were likely to suffer the fate of the Arrow. Success in the design of advanced missile and space systems ultimately depended on miniaturization, where size and weight (especially with spacecraft) had direct impacts on aerodynamics, fuel efficiency, and cost. By the end of the 1950s, military planners had realized that certain constraints – Canada's small population and its open and seasonal economy – made the creation of a domestic civilian expertise in transistor manufacture impossible. The only solution was for DND to invest in its own laboratories with the hope that national defence priorities would eventually translate into avenues of national innovation and welfare. Even so, investing in high-risk research and development that, as Vardalas states, "would have had little chance in finding support within an economy dominated by the export of natural resources," was truly an uphill battle.[50]

However, even within DND's own establishments, the lack of available technology created conflict and divisions regarding priorities. While the RCAF was assembling its proposed vision for the Space Defence Program, the DRB was simultaneously struggling to sustain the technology of its Alouette-ISIS project. As the launch of the first satellite in the series approached, the DRB barely had the computational assets or human resources to maintain its own space agenda, let alone another. Perhaps unknown to senior RCAF planners at the time, the Space Defence Program was probably technologically stillborn before it started. As demonstrated above, the evaporation of support for the RCAF endeavour in the mid-1960s perhaps merely served as a political rationalization of the technically obvious.

Conclusion

By the late 1960s, little if any corporate knowledge on outer space activities appeared to remain within DND senior ranks, and the idea that the needs of Canadian defence space could be met via US systems had already formed. A report on Canadian strategic intelligence requirements written in late 1968 noted that "at the present time there appears to be no unique Canadian military intelligence requirement that cannot be met by US satellites as long as they are willing to pass on the data. If arms control inspection techniques by satellites were to be perfected, Canada, as a highly developed middle power, might be assigned a monitoring task on behalf of the UN. The Canadian military and External Affairs would then become fully involved, however this is pure speculation at this time."[51] Similar attitudes prevailed with respect to other space ventures as the United States came to dominate the space sector in the West.

Unclassified DND archival documents record little continuing interest in the development of Canadian military space assets beyond 1969, other than simple monitoring and assessment. Although material held in national archives does indicate that DND-interdepartmental space committee meetings occurred during the early 1970s, these files remain classified and are largely inaccessible at this time. Unable to design and construct its own military space program during the 1970s, DND may have followed the example of Canada's civilian space program, which depended largely on the purchase of American-built commercial satellites and data. Thus, DND too may have purposely chosen to focus on the exploitation of space-derived data from allied sources. It is likely that, through various memorandums of understanding and intelligence-sharing arrangements with its allies throughout the 1970s, DND was able to secure some of the space data that it desired and later came to depend on.

It is difficult to pronounce on whether Canada's 1960s military space program was a success. Taking advantage of its expertise in post-war upper atmospheric research, the DRB was able to ensure limited Canadian access to space very early in the space race. This was of paramount importance considering that, like the United States, Canada needed the option of defending itself against space and as required through space. One would naturally expect DND to take advantage of its space access to quickly develop military space capability. Certainly, the United States welcomed Canadian contributions to the strategic defence of North America, and without some degree of space access or control, no defence space program would have been possible.

At first, Canada was very engaged in the process of militarizing and weaponizing space. Then, surprisingly, its military space leverage was subsequently squandered by the very organization responsible for military scientific and technological research and development. From an organizational point of view, it was simply incredible that the DRB was allowed to adopt and promote an agenda of "science for the sake of science" at the expense of science in support of military application. Yet timing was everything. When Paul Hellyer became minister of national defence, Ottawa was already looking to cut military spending, but the DRTE was well on its way toward launching Alouette. The project was also secure in the fact that it generated considerable national prestige and was probably deemed a worthwhile investment simply in political terms.

The RCAF Space Defence Program, however, born after Hellyer's arrival, faced insurmountable opposition from a financially constrained department about to undergo a major reorganization and a re-evaluation of its policy and objectives. Obviously wanting to protect its own interests, the DRB did little to promote other military space agendas, perhaps not realizing that it was ultimately contributing to its own demise. By undermining the development of other space initiatives within DND, the DRB effectively killed projects that could potentially have furthered its own goals.

Perhaps the DRB thought that its association with Alouette-ISIS would guarantee it a central role in a national space program. Instead, its shortsighted approach ensured that all military space enterprise was more easily stripped from DND when the time came to transfer national space activities to the civilian sector. With no loyalties or support in place, the DRB lost not only its space program but also the core of its space expertise – the DRTE – when the latter was transformed into the Communications Research Center within the newly created Department of Communications.

Another critical DRB error was the decision to wait for the official creation of a Canadian Space Agency before proceeding with its own policy and agenda. Arguing for the formation of such a body as early as 1962, the DRB fully expected that Ottawa would follow the American example by creating a civilian national agency out of its defence organization while still relying heavily on the latter for support. But Cabinet did not take this course, partially because of other priorities involved in the transformation of government during the 1960s and partially because the DRB already seemed to be functioning in that role. By focusing heavily on a scientific agenda at the expense of defence applications, it undermined any justification for creating what was perceived as a superfluous organization, which would simply duplicate what the DRB was already doing at a time when Cabinet wished to streamline and modernize government.

Systems management presented another considerable hurdle for Canada's military space program. Though certainly competent, DND lacked numerous opportunities to adequately research, develop, and sustain the robust technological infrastructure needed by such a complex program. Program management methodologies were virtually unheard of in the 1950s, but whereas mammoth projects forced the US missile and space community to adapt, no commensurate large-scale opportunities presented themselves to the RCAF to generate similar expertise.[52]

DND also faced substantial technological hardship in developing its own missile and space capabilities. Lacking support from the civilian industrial base or economy, it was forced to build, at great expense, most of the advanced electronics and computational services for its early programs. The defence budget freeze of the mid-1960s simply made it impossible to continue in this manner, and as a result, DND could neither develop new technologies nor retain the talented and experienced people required to build them. By the end of the decade, the DRB talent pool was decimated, as the majority of scientists and engineers had moved to Canada's civilian sector or to other projects outside the country.

Although, initially, DND had the strength to develop a military space program, numerous issues frustrated any hopes for this after 1970. Confronted with a rapidly mutating defence organization and increased political pressure on the scientific community and in industry, neither the DRB nor the RCAF had the mandate or the resources to fully exploit the military potential of space. When DND's limited resources were stripped in 1968 and 1969, especially in terms of staffing, its aspirations for its own space program were all but over for the time being. Still, without DND and the DRB, Canada's space program might not have evolved at all. In the late 1950s, no other organization was equally or better suited to exploit space or the opportunities being presented to Canada by the United States. As well, initially at least, DND had the capabilities and the human resources to meet the significant technological challenges of spaceflight, allowing it to achieve a successful satellite flight as early as 1962.

Of course, this view must be tempered by the *realpolitik* of the period. As political scientist Joel Sokolsky once noted,

> When it came to determining the size of a Canadian contribution this realism had a particularly Canadian flavour. Except in times of world war Canada's political leadership, has answered the question, "How much is enough," by posing another question, "How much is just enough?" How much is just enough to keep the CF [Canadian forces] roughly compatible with other allied forces? How much is just enough to demonstrate just enough support for allied calls for greater defence spending? How much is just enough to be able to make a contribution when Washington "rounds up the posse for its coalitions of the willing"? And how much is just enough to send? How much is just enough to maintain the close participation in North American aerospace defence?[53]

Applying the "just enough" principle, Canadian leaders have proven extremely adept at matching the limited real political benefits of defence spending with equally limited contributions, a tradition that began with continental defence and continues to the present in supporting allied missions overseas. As Sokolsky also suggested, "Washington seems to have adopted the 'Woody Allen' approach to Canadian (and other small allied) contributions, that 'ninety percent of life is just showing up.' In the interest of demonstrating that America can and does act multilaterally, the US government has historically welcomed whatever Canada could provide."[54] In return, Canada took an "easy rider" approach to allied commitments, and over time this became an essential part of its strategic culture. As noted by others, this stance has "guaranteed that Canada will always prefer to undertake less of an effort than its great-power partners want it to, but not so little as to be eliminated altogether from their strategic decision."[55]

Finally, there remains little doubt that the Conservative and Liberal governments immediately understood the implications of the militarization and weaponization of space, even though neither seemed to pay much attention to it. Henry Kissinger, the quintessential realist, appreciated Canada's diplomatic skill in maximizing its limited military commitments. As he put it, "Canada's somewhat aloof position combined with the high quality of its leadership gave it an influence out of proportion to its military contribution ... Canadian leaders had a narrow margin for maneuver that they utilized with extraordinary skill."[56] Until the advent of the flexible response option in the late 1960s, Ottawa fully accepted that it might have to militarize and weaponize space, just as it accepted nuclear, biological, and chemical weapons into its arsenal during the same period. For nearly a decade, the government and DND devoted significant resources to this end, a fact that presents a very different picture of the history of Canada's militarization and weaponization of outer space than previously known.

7

National Interests and a New Agenda for Outer Space

The findings and conclusions of the Chapman Report served as a catalyst for a new space agenda at the end of the 1960s. After years of aloofness and indecision, space advocates finally received clear signals from the government that Canada's role in space was to be formalized to some degree. In the fall of 1967, with an obvious emphasis on organization and communications, Cabinet directed the government to commit the resources necessary to build and deploy a domestic satellite communications capability by the early 1970s. This objective became the sole focus for almost all government space efforts during the next several years and was finally realized on 9 November 1972, with the launch of Anik-I, the world's first domestic communications satellite, into geosynchronous orbit. Anik also served as the mechanism through which Canada's space program made its transition from purely scientific and defence-oriented projects to those focused on commercial technological applications and industrial benefits. As well, it marked the end of both soldier and scientist-adviser dominance of Canada's space program as a generalist bureaucracy took control of space policy.

Also important to note, the launch of Anik-I did not have all the trappings of a sophisticated national space science program as was hoped by Chapman and his colleagues. Though Cabinet made a formal decision to pursue communications and published a white paper to that effect, it did not ratify an official national space policy until 1974, and the end product was designed to serve commercial rather than scientific agendas. Nor did Ottawa officially recognize Canada's space program through the creation of a Canadian Space Agency. Rather than invest heavily in new space infrastructure, Cabinet left the formulation of space policy to various advisory committees, and its satellite communications project proceeded under the direction of the more economical Department of Communications and a Crown corporation named Telesat Canada.

Space Detente

As the Cold War space race waned at the end of the 1960s, realism replaced idealism. Not only had the United States won the competition to place men in lunar orbit in December 1968, it soundly defeated the Soviet Union's lunar space program when Apollo XI landed on the moon in July 1969. Almost immediately after this climactic event, public interest in the romanticism of space travel appeared to recede, as did political support for expensive large-scale programs like Apollo. Public enthusiasm that had fuelled support for the space program at the height of the Cold War had been dulled by the American experience in Vietnam, and some began to question the need to beat the Soviets in space, given that they seemed to be taking the lead in military capability and influence on the ground.

Within a year of the first moon landing, the US government curbed funding and began diminishing the number of lunar missions. Initial plans had called for ten lunar landings, but Apollo XVIII, XIX, and XX were cancelled after cutbacks to NASA funding. In the end, seven were sent, and of these, only six reached the surface. Remaining resources were allocated to more functional earth-orbiting satellites and less expensive robotic interplanetary explorers.

Canada faced a similar trajectory in its own rocket and space program. The initial growth, spurred on by Cold War security concerns, defence, and a genuine attempt to seize new technologies, had reached a plateau by the beginning of the 1960s. The civilian-oriented industrial policy designed to take over from defence-led innovation was ill-defined and immature, and ultimately failed to sustain a high rate of technological development into the 1970s. This fact is demonstrated, for example, by efforts to create an indigenous computer industry during the 1960s. Though both government and private industry were urged to buy Canadian, the national economic policy of the day encouraged subcomponent manufacture and importation from the United States rather than the development of complete systems at home.

As well, the market for electronic goods and services simply did not exist in Canada to make large-scale technological development attractive or profitable. Attempts to use the government as a large "customer" also backfired when policy contradicted departmental directives. This was especially true with defence, which was designated as a prime advocate for advanced technological procurement while simultaneously having its budget and purchasing power drastically reduced.[1] Although there was little question that industrial competence in advanced technology, including space-related systems, was essential for the nation's future prosperity, the means simply did not exist to push Canada to sustain its own diversified and comprehensive rocket and space program.

The Liberal Cabinet chose to abandon those projects it felt were not beneficial, focusing instead on a select few areas where political and economic returns could be realized in the short term. Of those space capabilities easily within Canada's grasp, satellite communications became the priority. The country had achieved considerable success in the field thus far, had a small but highly trained and experienced cadre of scientists and engineers dedicated to it, and was capable of designing and manufacturing most of the technologies needed to put a communications capability into orbit. Any technologies and resources that Canada did not have could be procured from other countries such as the United States. In fact, the American industrial base became a solid resource. Since Canada could not sustain an indigenous space technology development, economic policy was designed around a type of moral suasion that encouraged American technology firms to cross Canadian tariff barriers and invest in high-quality research, development, and manufacturing in Canada to a level commensurate with their sales. This gave Canada access to the space technologies needed to build its own satellite systems.[2]

Satellite Communications

In the summer of 1967, with the results of the Chapman Report before it, the government moved to formalize its space agenda for the next several years. *A Space Program for Canada*, the Science Council's condensed version of the Chapman Report, had recommended the creation of a domestic communications satellite capability, and the Liberals chose to focus on this. Their decision made sense: in a country such as Canada, with its massive territory and small, widely dispersed population, survival and prosperity often depended on the establishment and maintenance of strong lines of communication. By the 1960s, television and telephone were rapidly gaining on radio as the primary medium of communication in Canada.[3] Possibly perceived as a modern-day version of the transcontinental railway, a satellite system that would link Canadians from coast to coast was endorsed by Prime Minister Pearson with little question.

On 6 June 1967, Cabinet met to instigate its short-term policy on satellites.[4] It decided to undertake immediate negotiations to register with the international community Canada's plans to establish a communications satellite system for domestic use. Part of this process would involve the creation of a special task force under the auspices of the Science Secretariat to undertake a comprehensive study of all questions on communications satellite development that the government should consider.[5] Members of the new Task Force on Satellites were to come from inside the government, and a list of potential candidates was approved in principle. MP Charles Mills "Bud" Drury, a decorated war veteran, was appointed chairman. A graduate of the Royal Military College of Canada, Drury had served with distinction during the Second World War, reaching the rank of brigadier. In 1949 he was appointed as deputy minister of national defence, a post he held until 1955. Elected and appointed to Cabinet in 1962, he was minister of defence production and subsequently the country's first minister of industry. Drury even served as Treasury Board president at a later date. He was

FIGURE 41 In the late 1960s, Charles Mills "Bud" Drury was the driving force behind Canada's transition to a national space strategy focused on telecommunications. National Research Council Archives.

more than suitable to head the task force, having knowledge and experience of scientific and technological matters from both military and civilian perspectives. Aside from Drury, other government members of the task force included Paul Martin, Mr. Winters, Jack Pickersgill, Paul Hellyer, Mr. Mellwraith, Judy LaMarsh, Mr. Sauvé, Mr. Benson, Pierre Trudeau, and Jean Chrétien.

Drury's Task Force on Satellites, which was to report back to Cabinet within the year, was authorized to commission special studies as required and retain any experts it needed to complete its work. Like Chapman's report, Drury's white paper was expected to spell out the way ahead for getting Canadian communication satellites into space. But unlike Chapman, Drury was subject to a sense of urgency from Cabinet. Perhaps hoping to establish a leadership role in the field, the government wished to get into the business before foreign competition degraded the market and diminished Canadian opportunities. It seemed that once the government had decided to pursue satellite communications, it wanted to proceed as quickly as events would allow.

The task force immediately embarked upon research and consultation. During the fall of 1967, it went to Italy, Britain, and France to exchange views and knowledge on satellite communications research. Interviews were also arranged with key personnel in Canadian government and industry. Similarly, the Air and Space Institute at McGill University's Faculty of Law was consulted, as were other academic and professional scientific organizations such as the Canadian Aeronautics and Space Institute. By early November, the task force had collected and analyzed enough material to allow it to submit an initial report on its findings and recommendations to Cabinet. Asked to deliver quickly, Drury obliged his employers.

His efforts were a success. Generally pleased with his findings and proposals, Cabinet accepted the task force's initial report in December 1967. Further consideration of the matter took place in early 1968, when Drury introduced a more finalized version of the white paper to Cabinet. Titled *White Paper on a Domestic Satellite Communications System for Canada*, the ninety-four-page bilingual white paper consisted of seven chapters and three appendices outlining current and future options for Canada in the field of satellite communications technology.[6]

The first half of the report summarized general techniques and achievements in satellite communications to date, including Canada's Alouette-ISIS project. The remaining four chapters approached the heart of the matter, most importantly providing a number of key decision recommendations to enable Canada to exploit domestic satellite communications by the early 1970s.

The task force envisioned a system of two synchronously positioned satellites in geosynchronous orbit over Canada, which would provide countrywide coverage. Each satellite would have the capacity to transmit between four and twelve television channels. Likewise, each channel could handle as many as six hundred two-way telephone circuits. One satellite would be the primary unit, whereas the other would play a backup role, acting as a redundancy for continuity of service should the primary unit fail or experience technical difficulties. A third satellite would be built and held in reserve on the ground, ready for launch should either of the other two fail because of a catastrophic event. Finally, the operational life of the satellites was expected to last between five and seven years.[7] Three major earth stations were envisioned, one each for television, telephone, and data. As well, a tracking, telemetry, and command facility would be needed to operate and maintain the satellites in their positions.

The three satellites were expected to cost between $40 and $75 million, including research and development. The ground stations would cost in the range of $100,000 each, with an additional few million dollars for the tracking, telemetry, and command station. However, like most space projects, this one had no precedent from which to extrapolate costs, which meant that accurately predicting its final price tag was difficult.[8] The white paper noted that the government was capable of assuming most of the financial outlay and that the expenses could be spread over a period of four to six years.

Perhaps most interesting was the report's brief discussion of the importance of the satellite project to the future of Canadian television broadcasting. In the late 1960s, many parts of Canada still had no access to broadcast television because of the prohibitive cost of placing terrestrial microwave feeds in remote areas. As a result, only those people living in large urban centres or along the Canada-US border where such systems were more common could easily and regularly receive television programs broadcast by satellite via local television stations. The report proposed to eliminate this problem through the use of satellites and locally placed receiving stations.[9] This was a small and yet remarkable demonstration of consideration for Canadian citizens and Native populations living in the North, and it showed that space technology could support national welfare if not the national economy.

Drury ultimately recommended that, because of its size and scope, the new satellite communications system should be a national project under government jurisdiction and control. He suggested the formation of a Crown corporation, which would ensure that the system was Canadian made, that new technologies would be infused into it, and that competition would occur between suppliers and contractors. Also, a Crown corporation could ensure an efficient sale of services while fulfilling "the minimum conditions for financial success."[10] With a certain amount of government control at the outset, Drury noted, Canada could count on some measure of regulation and profitability from the new technology.

Cabinet agreed in general with this assessment. It accepted the validity of shared public and private ownership of the corporation, given the magnitude of the cost, and it foresaw that the government would hold 51 percent of the corporation shares. Also, it agreed that the competition engendered between the new satellite system and the existing one would be good for Canada but that it "would eventually be such as to induce the private sector not to use the new system unless it was given some participation in its ownership and operation."[11] Similarly, Cabinet expected that the private sector would demand protection of its interests and cautioned that "it should be made clear that there would be no such protection in the future as the monopoly on the diffusion of ideas was very illiberal."[12]

The intent of the satellite effort was to link Canadians from coast to coast, preferably in both English and French. However, it was also seen as an opportunity to stimulate Canadian industrial capability and participation in advanced technology concepts, as by the end of the 1960s, Canada's backwardness in adopting modern space systems was starting to show. Although it had achieved success in space electronics and subsystem design, its experience in large-scale projects such as launch and satellite systems had been left to atrophy. Both government and industry were warning of the de-industrialization of Canada; the satellite project was designed to encourage research and development, regulation, industrialization, and even international cooperation and investment. Still, much work lay ahead before the satellites were put into orbit and the Canadian Broadcasting Corporation reached the country's living rooms.

Governmental Reorganization (Again)

Official organization remained the bane of Canada's space program throughout the 1960s. Without the oversight of a dedicated minister or space agency, setting goals, securing resources, completing projects, and maintaining a long-term vision were a constant struggle. Like all Canadian space projects, the satellite communications system needed both government and private resources. The difference this time, however, was the fact that Cabinet had endorsed the Drury White Paper.

In 1967 and 1968, the Department of National Defence remained the largest owner of national space resources and expertise. The Defence Research Board still oversaw the Churchill Research Range and its launch facilities and resources, and the Defence Research Telecommunications Establishment was preparing to launch the first of the ISIS series satellites. Though the National Research Council and various civilian universities boasted a reasonable cadre of space scientists, the engineering component rested largely within the domain of Canadian defence.

Rather than create parallel civilian organizations to implement the Drury White Paper plan, government proposed to transfer Canada's space expertise out of DND and into the civilian sector. This would facilitate civilian control over the country's space resources, while not demanding any significant degree of new funds or effort. Of course, DND was not pleased with the decision. It was then struggling to consolidate and ratify its own interests in outer space, and the loss of the DRTE and its associated space resources was a serious blow. Nevertheless, Minister of National Defence Leo Cadieux did not oppose the move, and thus DND could not stop the transfer. Satellite communications took precedence over all other space projects or efforts and was now Canada's national interest in space.

In April 1968, the government created the Department of Communications and endowed it with the oversight of all Canadian satellite communications development. Eric Kierans was appointed the first minister of communications. Soon afterward, the DRTE initiated the transfer of all of its resources and personnel from the DRB into the new department, where it was renamed the Communications Research Center.

The transfer of this group – the best concentration of experienced satellite engineers within the federal government – reflected both the end of the focus on defence- and scientific-oriented satellite projects and the rise of the new "civilian" space agenda. However, it also continued the trend of exclusivity in Canada's space program by isolating its components in various organizations rather than bringing them together under a single banner. The move did not go unnoticed, and once again Canada's primary space advocates attempted to convince Cabinet to bring the country's space efforts under the aegis of a single agency.

In July 1969, C.M. Drury, by that time chairman of the Privy Council Committee on Scientific and Industrial Research, tabled a confidential memorandum before his committee and later to Cabinet that outlined the imperative for a central body to coordinate Canada's new space activities.[13] After the repeated failures of the past, the moment was probably opportune to make such a proposal, and it demonstrated Drury's aggressive and insightful nature in these matters. He noted that the rapidly expanding space projects within government, the scientific community, and industry all continued to grow in relative isolation. Using the 1967 Chapman Report to bolster his argument, Drury explained that a centralized agency could better coordinate a national space program such as the domestic satellite communications system, and that it was the proper tool to provide oversight of the current transition from pure scientific research to space applications.[14] More importantly, Drury argued, amalgamating the existing organizations into a Canadian Space Agency would lead to savings in the space program budget.[15] Finally, his plan called for the immediate creation of an Interim Canadian Space Council (ICSC) composed of senior representatives from government, industry, and academia to prepare the groundwork for the official Canadian Space Agency.

Although Drury presented his arguments well, his efforts to galvanize support for a new organization were to no avail. The newly installed Trudeau Cabinet was not convinced of the need for a Canadian Space Agency, and no government department would support the plan on its own. Drury's approach to selling technological policy may have been partially responsible for this. His memorandums to Cabinet, which often exposed glaring defects in Canadian growth and development, warned of impending disaster if immediate action was not taken but then went on to advise caution in approaching the matter and to recommend resolving issues through a series of graduated steps. Examples of this are numerous. His 1969 memo on centralizing administration for Canada's space program followed this pattern, as did a 1968 memo to government outlining departmental policy regarding computer technology. Though undoubtedly well thought out, it did little to prompt decisive action and advance government technological agendas.

Also, some wondered whether another organization were needed. The recently established DOC was functioning well: during the previous year, it had successfully negotiated with the United States for purchase of a controlling stake in the INTELSAT I project, and it appeared well prepared for the upcoming creation of Telesat Canada in the autumn. Meanwhile, the NRC was already overseeing the majority of other Canadian space research. Neither it nor the DOC were particularly sympathetic regarding any proposal to create yet another administrative space organization, especially if this resulted in a diminishment of their own resources.

However, there was some consensus on the idea that existing federal agencies should better coordinate outer space activities. Therefore, this objective did receive support. After much consideration, the Privy Council Committee on Scientific and Industrial Research met once more and agreed that an Interdepartmental Committee consisting of representatives from the National Research Council, the Defence Research Board, and the Departments of Communications, External Affairs, Industry, Trade and Commerce, Transport, National Health and Welfare, Fisheries and Forestry and Agriculture, and Energy, Mines, and Resources be formed to consider three main objectives. The first was the desirability of establishing an Interim Canadian Space Council. If this were accepted, the committee would deal with the second objective – identifying the terms of reference, organization, and reporting channels of the proposed council. Third, it would designate a chairman of this Interdepartmental Committee based upon recommendations from the chairman of the Privy Council Committee on Scientific and Industrial Research.

The resulting organization was not exactly what Drury and his colleagues had desired, but it reflected Drury's staged approach to building policy and bureaucracy, and he appeared content to accept it for the present. As well, the concept of an interim organization made logistical sense, and it immediately suggested that, once the pieces were in place, the decision to create a Canadian Space Agency would eventually follow.

It is interesting that the Department of National Defence was not assigned to the ICSC; instead, the DRB was chosen to represent defence interests. Given the schism between the DRB and DND space agendas at the time, one wonders whether defence received any real representation within Cabinet, and it is very possible that its lack of direct voice or presence at the table simply helped to reduce interest in defence space programs at the end of the decade.

Finances also reflected the waning interest in defence space projects. In the fiscal year 1969-70, Ottawa spent roughly $17.3 million on space.[16] Of this, $7 million was budgeted for the NRC, with another $6 million allocated to the DOC. By contrast, only $4.2 million was assigned to DND, of which $2.3 million went to the Defence Research Establishment (Ottawa) and the remaining $1.9 million to the DRB. However, most of this funding was devoted to the ongoing ISIS satellite project with very little being distributed to missile or defence space projects. Finally, there was no indication that any new funding was planned for defence space proposals. For example, the 1971 *White Paper on Defence* had not referred to any DND space or missile project, yet another indication that these were no longer central to Canada's defence agenda. Clearly, government priorities lay elsewhere.

The formalization of the ICSC began in earnest in late November 1969, when the Cabinet Committee on Science Policy and Technology met to coordinate its organization. Again, time was a factor, as the United States had recently indicated that NASA representatives wished to visit Ottawa during December to discuss Canadian involvement in various American space projects proposed for completion in the next decade. Still, the country had no official body to coordinate Canadian space activities at the national or international level. In the end, it was NASA administrator Thomas Paine himself who visited Ottawa on 15 December 1969 and invited Canada to participate in the next era of American space travel and exploration. Quickly, the committee agreed that the ICSC – now renamed the Interdepartmental Committee on Space (ICS) – be established immediately with the same membership and terms of reference as outlined in Drury's July 1969 proposal to Cabinet.[17]

The ICS and the American Post-Apollo Program

The ICS, which met for the first time in January 1970, consisted of representatives from eight different departments. It was responsible for coordinating all government involvement in space programs, both present and planned. Its mandate "was to advise on policy and planning for Canadian space activities ... to ensure the coordinated development of government, university, and industrial activities, and international cooperation."[18] However, many within government felt that the ICS was unsuited to its task, lacking both the authority and the finances to determine clear objectives and make concrete decisions with respect to national interests in space. Some even argued that it "could lead to tragic consequences for Canada in the loss of technological opportunity" and that "a central coordinating and contracting agency for space research and development" was really required.[19] Succinctly, though no department would take on the space portfolio alone, many officials thought the ICS a poor substitute for a Canadian Space Agency and consequently gave it little respect or support.

Acting solely as a consultative body despite its policy-planning role and a Cabinet base, the ICS almost immediately (and unsurprisingly) ran into problems.[20] Many members of the loosely organized committee were inexperienced in outer space matters, indecisive about Canada's space policy, and divided regarding whether space activity was even an appropriate arena for Canadian foreign policy.[21] Meanwhile, other government departments quickly criticized the ICS for its general weakness in establishing national space policy and executing its decisions. The ICS had no funding authority, and its members were unwilling or unable to commit funds from their own departments. This, combined with a lack of political strength or consensus, meant that the ICS had no real authority and thus was ignored when it came to making serious decisions on Canada's major space projects.

Yet only when the ICS horribly mishandled a US suggestion that Canada participate in its post-Apollo program was its true weakness brought to light.[22] Shortly after he became president of the United States, Richard Nixon established a Space Task Group to advise the White House on US space program objectives in the post-Apollo period. The task group completed its report in 1969, recommending as a primary objective the development of "new systems and technology for space operations with emphasis upon the critical factors of (1) commonality, (2) reusability, and (3) economy."[23] At that time, NASA's main manned programs included the Apollo Applications Program (later called Skylab) and the Space Transportation System (or space shuttle). The task group also recommended that the United States promote international cooperation by inviting other nations to participate in its programs. Canada was specifically invited to do so but was also asked for an official reply as soon as possible.

The ICS was designated to coordinate the government's response to the invitation. It immediately established four subcommittees to address particular issues and produce reports to guide Cabinet on policy. The subcommittees were scientific research, space vehicles and propulsion, satellite applications, and international aspects. Each spent nearly a year investigating the American proposal and collecting data, after which all returned a generally favourable recommendation that the US offer should be accepted.

However, two particular issues of concern subsequently derailed the entire process. The first was that of international cooperation. In late 1970, C.M. Drury, then Treasury Board president, suggested in a public speech that linking international space activities with Canadian domestic priorities was desirable because it promoted Prime Minister Trudeau's vision for an independently minded foreign policy that was "the extension abroad of domestic priorities."[24] Though conceding that Canada-US space cooperation was "undoubtedly desirable and probably inevitable," he also asserted, "for this very reason there is a real political need to look beyond the continental relationships. Association with Europe offers such an opportunity and hopefully could be achieved at a tolerable cost."[25] Drury went on to suggest that Canada might even seek associate membership in the new European Space Agency then being formed to replace two existing institutions – the European Space Research Organization and the European Launcher Development Organization.

Intensifying his push for cooperation with Europe, Drury stated that Canada "would probably have more influence in the process of evolving an international institution" if it joined with Europe rather than maintaining the status quo and trying "to influence the United States on the strength of what would necessarily be a relatively very modest contribution to the overall NASA program."[26] He argued that entering into an association with Europe would give Canada access to new scientific and technological connections with countries such as West Germany and France, and would also provide commercial opportunities for the research and development gained through the US post-Apollo Program. Finally, he stated that Ottawa "intends to pursue a space policy consistent with Canadian resources and Canadian objectives."[27] Though all of this was valid to some degree, the undertones of Trudeau-era "third option" politics – that alternatives to Canada-US cooperation should be found – could not be ignored.

Though the scientist lobby in the ICS tended to oppose defence-related or exclusive Canada-US cooperation relationships, it virtually reversed this stance where space policy and programs were concerned. Few felt that space activities were the appropriate venue for practising foreign policy, but the fact remained that, in 1970, the United States was the only country other than the Soviet Union with a mature launch capability. Neither Europe nor Japan (Drury had also suggested cooperation with the latter) had assured access to space, and the European launch program was notorious for failure and disjointed management.[28] Some warned against becoming involved with Europe's space organizations, which lacked cohesion and a common sense of purpose and were constantly quarrelling among themselves.[29] Finally, as far as space activities were concerned, Canada had no history of cooperation with Europe or Japan.

By contrast, the United States had assured access to space, had some degree of space control, and had long been a partner with Canada in rocketry and space ventures. Some ICS members felt that diluting Canada's limited space resources by chasing unstable relationships with underdeveloped European organizations in order to satisfy a political agenda was irresponsible and wasteful.

The pro-Europe lobby stated that the US space program should not encompass the whole of Canada's future strategy and that the disarray in Europe's space planning was temporary. The anti-Europe faction disagreed and argued that little connected Europe's program to Canada's. Unable to reach a consensus, the ICS began to implode and ultimately the controversy prevented it from providing Cabinet with an appropriate response and agenda for participation in the US post-Apollo Program. By the time this issue was settled but not resolved, Canada had missed its opportunity to take part in the American Skylab space station project.

Another major problem confronting the ICS was that of financial resources. Any contribution to Skylab or the space shuttle had to be reviewed within the context of the Canadian aerospace program.[30] In 1970 most of the program was committed to developing the technology for vertical takeoff and landing and short takeoff and landing; there was no new funding for Canadian space projects beyond that already committed. How, then, could Canada afford to be involved? Cost sharing was out of the question: NASA's firm policy was that participating nations were financially responsible for their contributions and must pay their own development costs.

Taken together, the short timeline, the lack of an overall Canadian policy on space, financial concerns, and the congestion caused by the debate over European cooperation all contributed to the ICS's inability to formulate a government response to the American post-Apollo invitation. Though this failure was not entirely its own fault (the ICS was always perceived as an interim organization, and it was attempting to find scientific consensus during the height of the national science policy debate), it never fully recovered from this initial folly. Though the ICS existed for several years after this point, government decision making on science and technology shifted to other parties and organizations.

MOSST and Science versus Bureaucrat

National reviews of science policy, including that by the Senate Special Committee on Science Policy (the Lamontagne Committee), had nearly all recommended the creation of a science portfolio within government. The existing policy structures such as the ICS were not truly effective in advising Cabinet, and as political analyst Jocelyn Ghent suggests, the new ministry-of-state concept employed in other portfolios "appeared to fit the need for centralization of science policy efforts."[31] Thus, in October 1971, the Ministry of State for Science and Technology (MOSST) was created to oversee and coordinate science and technological areas, including Canada's space program.

MOSST was responsible for working with other departments to improve both the formulation and the execution of science and technology policy, especially that related to international cooperation. However, like the ICS, it too was solely an advisory and coordinating body without a formal decision-making mechanism or dedicated funding. What it did have was a clear mandate from Cabinet and a considerable degree of high-level political support. Still, MOSST found itself immediately challenged by veteran departments that already had considerable experience with international responsibility and were unwilling to yield it up to yet another new and unproven cog in the bureaucracy. In particular, the scientist and technologist advisers were reluctant to have their areas of expertise subjugated to more "generalist" bureaucrats whom they doubted could fully appreciate either the situation or the objective of Canada's science and technology programs. As James Hyndman noted in *International Journal,* scientific-oriented departments "tend[ed] to question the wisdom of turning over science and technology matters from the hand of the expert to the hand of the generalist, and they [were] also apprehensive of the interference of foreign policy goals with the rationality of the missions."[32]

That the relationship between scientists and bureaucrats should be acrimonious was nothing new, but it did underscore the difficult transition occurring in the overall relationship between science and government in Canada during the early 1970s.[33] In the decades immediately following the Second World War, the need for technologically enhanced security allowed both science and defence to play prominent roles in government decision making. By the late 1960s, that role had begun to change, and the standing of both scientist and soldier in policy making decreased noticeably. This occurred for many reasons. Some have argued that, though the elitism of scientific advice in government was warranted during the 1950s and early 1960s, it was wholly out of place by the 1970s.[34] Others offer more philosophical explanations: as the scholar Yaron Ezrahi suggests, "the Icarian dream of flying on toward a 'knowledgeable society' in which ideology and politics are replaced by technically rational choices approved by an informed public, may have lost its earlier hold upon the political imagination."[35]

In the case of Canada's transition, other factors were probably influential as well. Trudeau's style of government and political agenda differed strikingly from those of his predecessor, and science and technology were not always considered priorities. Social agendas were also overtaking security and industrialization interests in Cabinet, and funding traditionally reserved for technology and defence was diverted elsewhere. As well, both soldiers and scientist-advisers were increasingly disillusioned with their function in government. In 1970 neither the Defence Research Board nor the National Research Council had the same degree of access or influence in Cabinet that they had enjoyed only a decade before.

The creation of the Ministry of State for Science and Technology signalled that Canada's approach to international scientific cooperation would change further. Voicing an argument that some scientist-advisers may not have welcomed, government officials characterized the scientist lobby as lacking in appreciation of the larger political picture. As one contemporary political scientist writes, "They [the scientist-advisers] usually show little understanding for the argument that only ministries of foreign affairs have the necessary overview to evaluate the respective merits of competing priorities."[36] Although some departments felt that MOSST's activities tended to usurp their territory, Cabinet perceived the ministry as an important component in improving Canada's foreign policy decision making.[37]

Interestingly, the creation of MOSST was supposed to streamline policy and decision making in Canada's international science and technological matters, but at first it produced the opposite effect. The ministry initially had difficulty developing productive working relations with other government departments, as its function as "policy coordinator" seemed both threatening and unwarranted. At times, MOSST even appeared to bully its way into the missions and agendas of other departments. It has been suggested that MOSST's high public profile prompted it to address everything at once and to make hasty decisions without properly consulting the departments that were likely to be affected.[38] This behaviour "had the effect of diminishing the government's coordination of international science and technology cooperation instead of strengthening it as intended."[39]

It was in this manner that MOSST approached its coordination of the country's space program. Seizing on the ICS failure to properly address the American post-Apollo Program invitation, MOSST submitted a memorandum to Cabinet in August 1972 offering alternatives to the current coordination of activities in space technology.[40] Of course, all were based on the expectation that MOSST would assume control over this field. "Canada lacks an adequate mechanism for planning and coordinating the application of space technology in the national interest," the memo asserted, and suggested the following three options by way of remedy:

1 The ICS could be reactivated in its present form but report to Cabinet through the Minister of State for Science and Technology;

2 The committee could become advisory to MOSST and its responsibilities for formulating policy and coordinating activity could be assumed by the ministry; or,
3 MOSST could take over the responsibilities of the ICS and the ICS itself could be dissolved.

The minister initially advocated for the second option because MOSST lacked expertise in space technology and would need ICS advice as new space projects evolved.[41] This surprised some ICS members, who fully expected the demise of the committee once MOSST had set up a space task force of its own.[42]

The reality was that, difficult though it may have been for MOSST to accept, it needed the ICS to maintain oversight over current space activities and to help integrate these projects into a more formalized policy. For example, the nature and complexity of Canada's satellite communications project had increased tenfold when the United States announced plans to initiate its own program in 1970, and Ottawa chose an American rather than a Canadian prime contractor to build its communication satellites. This move, which was certainly not supportive of the "made in Canada for Canada" approach employed earlier to sell the government on domestic satellite communications, did not correspond with a space policy that was supposed to put Canadian interests first.

MOSST's initial failure to connect with other government bureaucracies seriously hindered space policy formulation throughout 1972 and 1973, and the overcommitment of its limited resources to maintaining a high public profile in all science and technology matters hampered efforts to produce an effective working document before the summer of 1973. MOSST needed to consult with ICS members and their respective departments, with the result that its guidelines and recommendations for a Canadian space policy were not presented to Cabinet until the spring of 1974.

In April the Liberal government published Canada's first official civilian space policy. Submitted by Minister of State for Science and Technology Jeanne Sauvé and developed with the full participation of the ICS representatives, the document finalized Canada's plan to emphasize space applications in preference to pure space science. Leaning heavily toward industry, the policy outlined four key points. First, the use of space applications had to contribute to established national goals. Second, the growth of the Canadian space industry had to be ensured by moving government space research into industry in accordance with a "make or buy" policy. ("Make or buy" was a term employed throughout Cabinet-level memorandums in defining how technology was to be acquired.) Third, Canada's satellite systems must be designed, developed, and constructed in Canadian industry, and finally, though it would rely on foreign launch services, the domestic space industry must be improved to meet needs at home. This last point was obviously included as a result of the very bad publicity sparked by the selection of an American company, Hughes, rather than RCA Victor of Montreal, to provide the satellites for the country's communications project.[43]

The policy made explicit what had been implicit since the mid-1960s – a transition from purely scientific research to space applications that contributed to the increase of national welfare. Satellite communications, then Canada's main space effort, clearly fit the bill. Yet the policy recognized that diversification was also required, and in this it reflected the Cabinet decision to expand Canada's space technology beyond ionosphere research and telecommunications.

More importantly, the make or buy policy, which encouraged joint government-industry endeavours to develop an indigenous productive capability, was aimed at satisfying domestic space system requirements, providing high technology employment opportunities, and enhancing the ability of Canada's aerospace industry to penetrate additional export markets, particularly those in the United States. Such an approach was also expected to enhance Canadian knowledge of space systems and to create opportunities to acquire new space systems hardware.

The caveat regarding foreign launches marked the continuation of Canada's dependency on international cooperation to achieve its space goals. Until the mid-1960s, Canada's space program was largely self-sufficient, with the exception of heavier launch capability. Though government had been advised to build and operate a launch vehicle pad at the Churchill Research Range for the American-designed Scout rocket, this plan never went beyond the concept stage, and Canada continued to rely on the United States for access to space. This became a real concern as growing numbers of nations sought access to launch services during this period, for additional customers meant increased competition for priority and preference. To ensure that Canada retained access throughout the 1970s, its space policy directed that it should consider participating in the supplying nation's space program (in this case, that of the United States) in return for guaranteed access to space. Fortunately, the mechanisms were already in place to allow that to happen.

Conclusion: The New Agenda for Space

The satellite communications project was initiated to organize and focus Canada's space efforts. In some regards, it was successful, producing the world's first domestic communications satellite system within five years and generating enough attention and impetus that the highest levels of government recognized some of the needs of Canada's space program. Nevertheless, it failed in other ways. Rather than prompting the creation of a space agency with a broad mandate for science and exploration, it secured a specialized and niche-oriented future that put commercial application permanently ahead of science. The transformation was so complete that Canada did not launch another scientific satellite of its own for almost thirty years.

The realization of the satellite communications agenda marked the formal termination of Canada's Cold War space defence programs, a course that was both necessary and inevitable. The space program had reached an impasse in 1972. The launch and satellite projects had experienced early triumphs, but the Cabinet decision to concentrate on semi-privatized satellite communications during the early 1970s greatly reduced Canada's overall space potential. Without a national policy and an effective centralized authority, neither defence nor the civilian sector was able to advance its own interests on a broad scale let alone take part in further international efforts. Realizing its own paralysis, MOSST sought a plan that harnessed the assets of the country's existing aerospace industry and combined it with a political and economic agenda for expansion. This in turn suggested a plan for sustained international cooperation. Therefore, the 1974 space policy focused on this theme.

Changes in the emphasis of the Canadian space program also formed the relationship between science, defence, and government. In the 1960s, the link between defence and government had fundamentally altered in connection with space policy development, a pattern that repeated itself with scientist and government during the policy formulation of the 1970s. Though scientist-advisers played a central role in initiating the Canadian satellite communications project, their influence on decision making did not persist to the point where Canada's official space policy began to take shape. As explored throughout this chapter, the machinery of government solidified in the latter days of the national science policy debate so that, in the end, it was the bureaucracy, not the science and technology community, that controlled the final stages of space policy development. As well, social and economic goals rather than purely scientific objectives lay at the heart of the plan. The 1974 space policy was tailored to foster much-needed advanced industrial development. It was also clearly designed to serve specific national interests.

Conclusion

DURING THE DECADES AFTER the Second World War, conceptual, design, and technological evolution in flight continued unabated as humankind pushed the boundaries of its own reach. Fuelled by the need for security and an unquenchable thirst for knowledge, the ascent into outer space was, it seemed, inevitable.

The Second World War expansion of science and technology put Canada in a position to capitalize on its Cold War geo-strategic location between the Soviet Union and the United States so as to gratify its immediate needs and desires. When space travel became a reality, Canada possessed the means with which to take part, but technology itself would not determine the extent of its involvement; nor would its participation be defined solely by economics or politics. Instead, the interaction between the two determined the nature of the Canadian program and explains the often disjointed and sporadic character of its achievements.

Undoubtedly, national security and military enterprise fuelled Canada's developments in rocketry and upper atmospheric research during the first post-war decade. Striving for self-sufficiency and self-identity in Cold War politics sustained technological development at first, with the expectation that military technologies would eventually be transferred to the civilian economy. In the United States and Britain, national defence had contributed to national socio-economic welfare during peacetime; Ottawa expected the same process to occur in Canada.

When Canada's space activity moved from rocketry to satellites, three communities exerted influence on its direction – science and technology, defence, and government. When the three interacted, differing philosophies and conflicts arose. Though Canada's space program was influenced and altered as a result, it neither disappeared nor declined technologically as some have suggested. Canada still demonstrated considerable achievement in outer space but always in true Canadian style – "just enough" to benefit directly from the investment made.

Certain technological opportunities were lost. DND constantly struggled to acquire the new technologies needed for missile defence and the exploitation of space and thus was unable to accomplish specific military aims or to adequately promote its defence space agenda within government. That said, however, the defence community was attacked by its own scientific community, which subscribed to a different space agenda and was plagued by contradiction and sometimes hypocrisy from the political level.

In the 1960s, after Ottawa had failed to sustain defence spending and procurement during the late 1950s, the Royal Commission on Government Organization argued that "the [defence] procurement effort must be guided by adequate appreciation of the present and future potentials of the economy in fields of research, development, and technology which can, by proper stimulation and support, underpin the country's economic strength and potential economic growth."[1] Why place so much expectation on the Department of Defence Production and DND when government funding of these institutions had decreased so sharply during this period? Similarly, the commission stated that "the Canadian defence establishment marked a lack of confidence in Canadian science and technology as a whole," which seems very hypocritical given that DND research laboratories designed and built Canada's first computer, rocket, and satellite.[2]

Scientific and engineering opportunities were also lost. The exclusive attention on satellite communications ended research and development on next-generation launch systems as well as early space systems for remote sensing, earth resources, and aeronautical research. Science was forced off of Canada's space agenda by labour-force development and economic and industrial benefits and would not be officially reinstated until the 1980s.

Still, the space program did achieve its victories despite (or because of) government direction and intervention that affected technological development. The defence establishment carried space development through the first three decades of the Cold War to the point where the civilian bureaucracy assumed control of its future. Science played a role throughout, initially a large one but later limited to bolstering the needs of commercial space application. In the end, the government applied its typically shrewd methods and got what it needed – international prestige, a domestic communications system, a strong technological relationship with the United States, and an invitation to future space activities as they arose.

Finally, the traditions formed during the first three decades of Canada's space program have survived to the present day. The deployment of the Canadarm II to the International Space Station by Canadian astronaut Chris Hadfield and the crew of the space shuttle Endeavour publicly re-emphasized the high level of cooperation and integration that Canada has achieved with both the United States and other international partners. At first glance, Canada's presence as one of the five key players in the space station venture suggests a remarkably capable program with significant resources to devote to large-scale activities. Whereas the former may be true, the latter is most certainly not: the annual program budget, of approximately $334 million, is comparatively small. Thus, Canada's prestigious membership within the space station community is more the result of a savvy tradition of strategy, policy, and diplomacy rather than major financial effort.

Examining how Canada's space program arrived at this point, exploiting policy in lieu of financial assets or material resources, is crucial to understanding that what initially appears rather lacklustre actually constitutes a remarkable accomplishment. With very limited resources, Canada's space program has not only survived, it has also managed to prosper as an example of national achievement. It did not evolve in a vacuum: it responded to the indigenous and international influences that fuelled the Cold War space race. The United States played a major role in supporting and encouraging it but did not exert the controlling or coercive influence often witnessed in bilateral defence or trade policy.

The Cold War shaped Canada in every conceivable way. From the perspective of security, the country's relationships as well as the scientific and technological expertise of its defence community ensured that Ottawa remained positioned advantageously to exploit new opportunities. When humanity reached forth under the guise of defence into space, Canada was able to join that new adventure.

The nature, characteristics, and evolution of science, technology, and society in Canada, especially after 1945, requires more investigation than the scope of this book allows. Nevertheless, this examination of Canada's military space program between 1945 and 1974 presents yet another comprehensive case study of how Canada strove to create a national technological competence during the height of the Cold War era. It is from such studies that we might gain not only a broader understanding of how science and technology influenced defence, but also of how Canada's approach to defence problem solving at times influenced and shaped the very nature of the Cold War itself.

Appendix
Members of the Associate Committee on Space Research, 1959-63

Chairman

D.C. Rose, Division of Pure Physics, National Research Council, Ottawa.

Secretary

B.D. Leddy, Division of Administration and Awards, National Research Council, Ottawa.

Members

J. Auer, Medical Research Council, Ottawa.

J.H. Chapman, Radio Physics Laboratory, Defence Research Board, Ottawa.

R.F. Chinnick, Defence Research Board, Quebec City.

J.W. Cox, Directorate of Physical Research, Defence Research Board, Ottawa.

P.A. Forsyth, Department of Physics, University of Western Ontario, London.

C. Fremont, Department of Physics, Laval University, Montreal.

G.M. Griffiths, Department of Physics, University of British Columbia, Vancouver.

A. Kavadas, Department of Physics, Dalhousie University, Halifax.

D.P. McIntyre, Air Services, Meteorological Division, Department of Transport, Toronto.

D.W.R. McKinley, Radio and Electrical Engineering Division, National Research Council, Ottawa.

G.S. Murray, United Nations Division, Department of External Affairs, Ottawa.

R.W. Nicholls, Department of Physics, University of Western Ontario, London.

G.N. Patterson, Institute of Aerophysics, University of Toronto, Toronto.

H.I. Schiff, Department of Chemistry, McGill University, Montreal.

M.M. Thomson, Dominion Observatory, Department of Mines and Technical Surveys, Ottawa.

F.R. Thurston, National Aeronautical Establishment, National Research Council, Ottawa.

H.J. Williamson, Telecommunications Branch, Department of Transport, Ottawa.

B.G. Wilson, Department of Physics, University of Alberta, Calgary.

A Note on Sources

Like other nations and allies, Canada joined the Cold War race for space. Yet there has been little examination of the history of Canadian science and technology or in particular of Canada's space program, which depends upon technology. Aside from the infrequent government-produced general overviews and technical summaries, only a few academic chapters and articles have been written on the subject by a handful of journalists and scholars. This may seem odd, given the popularity of space travel and discovery. Yet the field has not attracted a large number of Canadian political scientists or historians.

Most of the available secondary literature may be classified as purely technical narrative. Historians who have attempted to examine both the design choices and the socio-economic factors that stimulated and shaped the process of space exploration often misidentify the determinants of change and fall into either an "internalist" or "externalist" historical narrative. The former treats technological change as transcendent of human will or intervention. The latter asserts that it is affected only by social, economic, or political factors. *Defence and Discovery* has followed a methodology recently proposed by German historian H.D. Hellige, which holds that the process of technological change results from the interplay between cultures, economics, and politics. According to this model, technological change does not occur in a vacuum: rather, it is shaped by the nature of the society (in this case Cold War Canada) within which science, defence, and politics are practised.[1]

American military and computer historian Alex Roland once wrote that "documenting the history of modern technological development poses challenging problems and exhilarating opportunities."[2] Although he was referring to a subject covering the years 1983 to 1993, his comment was certainly germane to my own experience in completing this history, though I might add that the difficulties for scholars in this field will continue to outweigh the opportunities until further study is devoted to the general history of science and technology in Canada.

Serious obstacles to my research appeared at the outset. First, I lacked a distinct starting point for the subject. Perhaps fewer than six books have been written on the entire fifty-year history of Canada's rocket and space program; half of these are government-sponsored media, and none are academic in nature. There are no official histories of specific Canadian programs except for the Hermes CTS satellite, which is covered in some detail in *Spacebound* (1982), by T.R. Hartz and I. Paghis. Nor were any detailed quantitative analyses of Canada's space program completed after the 1967 Chapman Report. Similarly, none of the major personalities who shaped Canada's space legacy have received full-length biographical treatment, not even John Chapman, popularly known as the father of the Canadian space program and for whom the modern Canadian Space Agency is named. The works that do discuss Chapman have accorded his life and contribution no more than a cursory look. Essentially, a search for secondary source information on Canada's rocket, missile, and space programs during the period of this study (1945-74) turned up nothing beyond the odd and often erroneous paragraph. More recently, a handful of suborganizations such as Canada's Communications Research Center have taken the initiative to encourage the volunteer development of their own online historical collections, but these efforts still remain too few at present to fill the tremendous gaps that exist in this field of study. Though the United States and other nations have chronicled their space history in considerable detail, Canada's particular achievements in outer space are notable only by their absence from the public record.

Obtaining useful primary source evidence for this book was always a privilege rather than a right. Unlike my colleagues, who often completed their research at one or two archives, I found my sources literally scattered across more than fifty agencies, institutions, companies, and establishments in North America and elsewhere. In Canada, significant government-related sources were located in the Library and Archives Canada Record Group 24 (National Defence), 25 (External Affairs), and the accessioned files of the Defence Research Board and related agencies. As well, various files at the Directorate of History and Heritage were relevant to Chapters 4, 5, and 6. Files concerning the National Research Council's involvement in outer space research and development were available at Library and Archives Canada and the NRC itself, with space-related materials surfacing in the records of half a dozen other government departments, depending on the particular subject and period.

Though many government files germane to my study did exist, almost none had been vetted for release to the public, which meant nearly three years of submitting access to information requests to get these much-needed sources declassified. Still, not all government documents were releasable, particularly those associated with national security and defence. The defence origins of the Canadian rocket and space program, and its cooperation with the defence communities of the United States and the United Kingdom, inevitably meant that some information would remain classified, as it has not yet been released in other countries, or the government is unable to disclose information shared from other sources. As well, the nature of certain relationships remains classified, even though nearly half a century has elapsed since the time in question. Because of its subject matter, some information remains classified regardless of its age. In such cases, I employed alternative sources wherever possible to make inferences regarding the likely course of events. Obviously, more detailed scholarship awaits this subject in the future. Finally, the changing security environment after the events of 9/11 meant a general slowdown in response to requests for information on all manner of subjects related to missile, rocketry, and defence space programs, for obvious reasons. These circumstances created challenges on more than one occasion.

Other avenues proved less helpful than expected. Canada's own space agency did not exist prior to 1989; therefore, no central repository or historical office was responsible for collecting and preserving materials related to Canada's rocketry and space programs, let alone the drafting of historical records and manuscripts (and such an office has yet to be formed). Instead, documents were retained, if they were retained, by the various departments, interdepartmental committees, or establishments involved in a given project or program. Although much of this material eventually found its way to Library and Archives Canada, some of it remained in the original institution or sometimes with whatever other organization was willing to accept the "obsolete" papers. As times changed and organizations were restructured, valuable records were often abandoned or deliberately destroyed.

In some instances, this situation proved nearly disastrous, especially with private-sector bodies involved in various aspects of Canadian rocket and satellite development. For example, the companies that descended from those who pioneered the Black Brant program had abandoned or destroyed any documents that were more than fifteen years old and not required for legal purposes. As a result, most of the early material on the Black Brant program was gone forever. That which remained generally lacked any finding aid, administrative history, or descriptive information. Likewise, thousands of photographs remain uncatalogued and without any detailed information on what the pictures are or who might be in them.

Over the course of the development of this book, always keen to uncover any new evidence, I did locate both documents and photographs in a variety of unlikely places ranging from private libraries to garage sales and flea markets, and even on Internet auction sites. Much of this research is currently being turned into a fond so that other scholars may benefit from it in the future.

Therefore, scholars who enter this field of study should do so with cautious expectations at present, keeping in mind, however, that it is only through sustained effort and the revealing of more archival evidence that the general situation will improve for all who are interested in the topic over the longer term.

Notes

Introduction

1 General histories include T.A. Heppenheimer, *Countdown: A History of Spaceflight* (New York: J. Wiley and Sons, 1997); Andrew Chaikin, *A Man on the Moon* (New York: Penguin Books, 1994); and Asif A. Siddiqi, *Challenge to Apollo: The Soviet Union and the Space Race, 1945-1974* (Washington, DC: NASA, 2000).

2 J.J. Brown, *Ideas in Exile: A History of Canadian Invention* (Toronto: McClelland and Stewart, 1967), 339-40.

3 Christian DeBresson, "Have Canadians Failed to Innovate? The Brown Thesis Revisited," *History of Science and Technology in Canada Bulletin* 6 (January 1982): 10-23.

4 Merritt Roe Smith, ed., *Military Enterprise and Technological Change* (Cambridge, MA: MIT Press, 1987), and C.N. Hill, *A Vertical Empire: The History of the UK Rocket and Space Programme, 1950-1971* (London: Imperial College Press, 2001).

5 For a detailed study of the origins of the NRC to about 1935, see Mel Thistle, *The Inner Ring: The Early History of the National Research Council of Canada* (Toronto: University of Toronto Press, 1966).

6 For a thoroughly detailed analysis of Canadian wartime defence science, see Donald H. Avery, *The Science of War: Canadian Scientists and Allied Military Technology during the Second World War* (Toronto: University of Toronto Press, 1998), and John Bryden, *Deadly Allies: Canada's Secret War, 1937-1947* (Toronto: McClelland and Stewart, 1989).

7 Canadian official documents suggest that Canada was the third country to go into space after the Soviet Union and the United States. This ranking has been challenged by other early spacefaring nations and remains a matter of debate.

8 Douglas Owram, *Canadian History: A Reader's Guide* (Toronto: University of Toronto Press, 1994), and Carl Berger, *The Writing of Canadian History: Aspects of English-Canadian Historical Writing since 1990* (Toronto: University of Toronto Press, 1993), 212.

9 An excellent history on the subject of Canadians in the American space program is Chris Gainor, *Arrows to the Moon: AVRO's Engineers and the Space Race* (Burlington: Apogée Books, 2001). By the same author, see also "Owen E. Maynard: Lunar Module Pioneer from Canada," *Quest: The History of Spaceflight Quarterly* 9, 3 (2002): 22-28.

10 The author has undertaken a small series of official historical analyses for the Canadian Space Agency's internal use. They have yet to be made available to a wider public.

Chapter 1: Cold War Security and Aerospace Defence Research Prior to Sputnik

1 For an overview, see Richard A. Jarrell, *The Cold Light of Dawn: A History of Canadian Astronomy* (Toronto: University of Toronto Press, 1988).

2 See Doris Jelly, *Canada: 25 Years in Space* (Ottawa: National Museum of Science and Technology, 1988), 13-17, and A.M. Pennie, comp., *Defence Research Northern Laboratory, 1947-1965*, Report No. DR179 (Ottawa: Defence Research Board, 1966), 5-11.

3 Donald H. Avery, *The Science of War: Canadian Scientists and Allied Military Technology during the Second World War* (Toronto: University of Toronto Press, 1998); see also J.H.

Chapman et al., *Upper Atmosphere and Space Programs in Canada: Special Study No. 1* (Ottawa: Science Secretariat, 1967), which is also published in abridged form as Canada, Science Council of Canada, *A Space Program for Canada,* Report No. 1 (Ottawa: Queen's Printer, Science Council of Canada, 1967).

4 J.J. Green. "Science and Defence," *Canadian Aeronautical Journal,* April 1955, 3.

5 D.J. Goodspeed, *A History of the Defence Research Board of Canada* (Ottawa: Queen's Printer, 1958), 6; see also M. Girard, "The Commission of Conservation as a Forerunner to the National Research Council, 1909-1921," in *Building Canadian Science: The Role of the National Research Council,* ed. Richard A. Jarrell and Yves Gingras (Ottawa: Canadian Science and Technology Historical Association, 1991), 19-40.

6 Philip C. Enros, "'The Onery Council of Scientific and Industrial Pretence': Universities and the Early NRC's Plans for Industrial Research," in Jarrell and Gingras, ibid., 41-52.

7 Yves Gingras, *Physics and the Rise of Scientific Research in Canada* (Montreal and Kingston: McGill-Queen's University Press, 1991). For this period, see also W. Eggleston, *National Research in Canada: The NRC, 1916-1966* (Toronto: Clarke Irwin, 1978).

8 A. King et al., *Reviews of National Science Policy: Canada* (Paris: Organisation for Economic Co-operation and Development, 1969), 43. See also Canada, Department of Reconstruction and Supply, *Research and Scientific Activity, Canadian Federal Expenditures, 1938-1946* (Ottawa: King's Printer, 1947).

9 Canada, Department of Reconstruction and Supply, *Research and Scientific Activity,* 42-45.

10 The relationship between Churchill and Lindemann is described in C.P. Snow, *Science and Government* (Cambridge, MA: Harvard University Press, 1961). For details on the relationship between Howe and Mackenzie and the rise of scientific influence in wartime Canada, see G.B. Doern, *Science and Politics in Canada* (Montreal and Kingston: McGill-Queen's University Press, 1972), 4-49.

11 Canada, National Research Council, *Annual Report: 1944-45* (Ottawa: King's Printer, 1945), 7.

12 C.J. Mackenzie, "Industrial Research in Post-War Canada" (paper presented to the Engineering Institute of Canada, Quebec City, 11 February 1944), cited in Goodspeed, *A History of the Defence Research Board.*

13 Goodspeed, ibid., 12.

14 "Cabinet Conclusion – Committee on Research for Defence," 3 October 1944, Privy Council Office Papers (PCO), Record Group (RG) 2, series A-5-a, 2636, 2, Library and Archives Canada (LAC), Ottawa.

15 Goodspeed, *A History of the Defence Research Board,* 13-16.

16 Charles Foulkes C.B., C.B.E., D.S.O. (1903-69).

17 Goodspeed, *A History of the Defence Research Board,* 19-22.

18 Ibid., 21.

19 Avery, *The Science of War,* 228-55.

20 Ibid.

21 Ibid., 28-41; see also A.M. Fordyce, "How It All Started: The Goforth Paper," *Canadian Defence Quarterly* 1, 4 (Spring 1972): 15-16.

22 A complete list of those present is available in the minutes of the first meeting of the Cabinet Committee on Research for Defence, 4 December 1945, also cited in Goodspeed, *A History of the Defence Research Board,* 41.

23 At the time of writing, there was no dedicated biography of Solandt, but biographical details may be found in several publications. See C.E. Law, G.R. Lindsey, and D.M. Grenville, eds., *Perspectives in Science and Technology: The Legacy of Omond Solandt* (Kingston: Queen's Quarterly, 1994); Goodspeed, *A History of the Defence Research Board,* 45-46;

George R. Lindsey, ed., *No Day Long Enough: Canadian Science in World War II* (Toronto: Canadian Institute of Strategic Studies, 1997), 264-66; and C.E. Law, "Omond McKillop Solandt: Operational Research Pioneer," *INFOR* 31, 2 (May 1993): 9-11. Solandt's personal papers reside in the University of Toronto Archives.

24 Soviet espionage in the West is covered in numerous volumes. Of particular quality and interest to the topic discussed here is Christopher Andrew and Vasili Mitrokhin, *The Mitrokhin Archive: The KGB in Europe and the West* (London: Penguin Press, 1999 and 2000).

25 For early Cold War Soviet espionage in the United States, see Alexsander Feklisov, *The Man behind the Rosenbergs* (New York: Enigma Books, 2001); Ronald Radosh and Joyce Milton, *The Rosenberg File* (New Haven: Yale University Press, 1997); John Earl Haynes and Harvey Klehr, *Venona: Decoding Soviet Espionage in America* (New Haven: Yale University Press, 1999); and Andrew and Mitrokhin, *The Mitrokhin Archive.*

26 J.L. Granatstein, *A Man of Influence: Norman A. Robertson and Canadian Statecraft, 1929-1968* (Toronto: Deneau, 1981), 172.

27 William Lyon Mackenzie King Diary, 7 September 1945, quoted in ibid.

28 Granatstein, *A Man of Influence,* 173-74.

29 Avery, *The Science of War,* 228-34.

30 Reg Whitaker and Gary Marcuse, *Cold War Canada: The Making of a National Insecurity State, 1945-1957* (Toronto: University of Toronto Press, 1994); J.L. Granatstein and D. Stafford, *Spy Wars: Espionage and Canada from Gouzenko to Glasnost* (Toronto: Key Porter, 1990); and J. Sawatsky, *Men in the Shadows: The RCMP Security Service* (Toronto: Doubleday Canada, 1980).

31 Avery, *The Science of War,* 228, and Granatstein, *A Man of Influence,* 174.

32 J. Starnes, *Closely Guarded: A Life in Canadian Security and Intelligence* (Toronto: University of Toronto Press, 1998), 76-81.

33 For a dated history, see Goodspeed, *A History of the Defence Research Board,* 47-49.

34 Ibid., 49.

35 Ibid.

36 The RCAF post-war agenda is surveyed in Randall Wakelam, "Flights of Fancy: RCAF Fighter Procurement, 1945-1954" (master's thesis, Royal Military College of Canada, Kingston, 1997).

37 Chiefs of staff meeting, 7 May 1946, cited in ibid., 49. See also O.M. Solandt, "Policy and Plans for Defence Research in Canada" (policy paper prepared for Cabinet Committee on Defence, May 1946), Part II.

38 "Defence Research and Development – External Policy, Organization of Research for Defence, and Defence Research; Organization of Board," 1946-47, PCO, RG 2, series A-5-a, vols. 2638, 2639, 2640, LAC. The Cabinet decision papers summarize more lengthy memorandums and reports submitted by the director general for defence research and the DRB to the prime minister for review and recommendation.

39 Senate of Canada, *Committee on Science Policy,* No. 9 (1969), 1264-65. Also quoted in Doern, *Science and Politics,* 92.

40 Goodspeed, *A History of the Defence Research Board,* 50-51.

41 Ibid., 51n10, 58-61.

42 Greig Stewart, *Shutting Down the National Dream: A.V. Roe and the Tragedy of the Avro Arrow* (Toronto: McGraw-Hill Ryerson, 1997), 62.

43 The early post-war RCAF fighter research agenda is examined in Donald C. Story and Russell Isinger, "The Origins of the Cancellation of Canada's Avro CF-105 Arrow Fighter Program: A Failure of Strategy," *Journal of Strategic Studies* 30, 6 (December 2007): 1025-50.

44 See Homer E. Newell, *Beyond the Atmosphere: Early Years of Space Science,* NASA SP-4211 (Washington, DC: NASA, 1974).
45 The final process to create the DRB is well covered in Goodspeed, *A History of the Defence Research Board,* 62-68.
46 Ibid., 109-10. General information on Canada's early post-war scientific research and development expenditures may also be found in King et al., *Reviews of National Science Policy.*
47 A. Chapnick, "Principle for Profit: The Functional Principle and the Development of Canadian Foreign Policy, 1943-1947," *Journal of Canadian Studies* 37, 2 (Summer 2002): 68-85, and A. Chapnick, "The Canadian Middle Power Myth," *International Journal* 55, 2 (Spring 2000): 188-206. See also Andrew Richter, *The Evolution of Strategic Thinking at the Canadian Department of National Defence, 1950-1960,* Occasional Paper 38 (Toronto: Center for International and Strategic Studies, York University, 1996), 11-35.
48 Defence Research Board, "Annual Reports," 1947-49, RG 19 520, file 124-62, pt. 1, LAC, and Defence Research Board, minutes of meetings, 1947-51, RG 24, series E-1-b 5250, file 19-73-4, LAC.
49 T.R. Hartz and I. Paghis, *Spacebound* (Ottawa: Department of Communications, Minister of Supply and Services Canada, 1982), 31.
50 Ibid., 49.
51 Ibid., 52.
52 Goodspeed, *A History of the Defence Research Board,* 192.
53 Ibid., 194.
54 G.D. Watson, "The Scientific Exploration of Space," *Canadian Aeronautical Journal* 6, 3 (March 1960): 87-88.
55 Goodspeed, *A History of the Defence Research Board,* 195-97.
56 "Biography – General," 1, John H. Chapman Papers (JHC Papers), Manuscript Group (MG) 31, series J 43, LAC.
57 Goodspeed, *A History of the Defence Research Board,* 197.
58 Sidney Chapman, *IGY: Year of Discovery* (Ann Arbor: University of Michigan Press, 1959); Marcel Nicolet, "Historical Aspects of the IGY," *Eos Transactions* 64, 19 (10 May 1983): 12-19; and Newell, *Beyond the Atmosphere.*
59 Some academic debate remains regarding whether Chapman or Berkner suggested the third IPY. For views on the former man's claim, see Chapman, *IGY: Year of Discovery;* the latter's claim is supported in Allan A. Needell, *Science, Cold War and the American State: Lloyd V. Berkner and the Balance of Professional Ideals* (Washington, DC: National Air and Space Museum, 2000).
60 L.V. Berkner, *Science and Foreign Relations* (Washington, DC: Department of State, 1950), quoted in Needell, ibid., 299.
61 G.K. Megerian, Secretary, V-2 Upper Atmosphere Research Panel, Minutes of the Rocket and Satellite Research Panel, reports 36-40, 7 October 1953 through 3 February 1955, NASA History Office Files, cited in Newell, *Beyond the Atmosphere,* Chap. 4.
62 Coordinating Committee for the Canadian I.G.Y. Program, notice of meeting, 28 October 1958, MG 30, series J 43 9, file 9-2, LAC.
63 Jarrell, *The Cold Light of Dawn,* 156-57.
64 "Speaking Notes – The International Geophysical Year Program in Canada," 1957, JHC Papers, MG 31, J 43 9, file 9-2, LAC.
65 Watson, "The Scientific Exploration of Space," 87-88.
66 H.B. Lutz, "Preparations for the IGY 1956-57," in Pennie, *Defence Research Northern Laboratory,* 84.

Chapter 2: Missiles, Rocketry, and the Development of the Black Brant Launcher

1 Royal Canadian Air Force Guided Missiles Committee, "RCAF Participation in Guided Misslry [sic]" 7 May 1949, (secret) memorandum, Record Group (RG) 24, vol. 5429, file HQS 446-23-13, Library and Archives Canada (LAC), Ottawa.

2 For very brief discussions of Canada's rocketry program, see T.R. Hartz and I. Paghis, *Spacebound* (Ottawa: Department of Communications, Minister of Supply and Services Canada, 1982), and Doris Jelly, *Canada: 25 Years in Space* (Ottawa: National Museum of Science and Technology, 1988). The best "official" history of the Black Brant remains a single chapter in the second volume of the Bristol Aerospace company history. See *Bristol Aerospace Limited: 50 Years of Technology, 1930-1980*, vol. 2, *The Second Quarter Century* (Winnipeg: Bristol Aerospace, 1985).

3 *Bristol Aerospace Limited*, 31-32.

4 Ernest A. LeSueur, "Rocketeers," *Canadian Defence Quarterly* 9, 3 (April 1932): 374.

5 Mel Thistle, *The Inner Ring: The Early History of the National Research Council of Canada* (Toronto: University of Toronto Press, 1966), 345-49.

6 For studies on science in Canada between 1880 and 1945, see Yves Gingras, *Physics and the Rise of Scientific Research in Canada* (Montreal and Kingston: McGill-Queen's University Press, 1991), and Richard A. Jarrell and Yves Gingras, eds., *Building Canadian Science: The Role of the National Research Council* (Ottawa: Canadian Science and Technology Historical Association, 1991).

7 D.J. Goodspeed, *A History of the Defence Research Board of Canada* (Ottawa: Queen's Printer, 1958), 127-33.

8 J.H. Chapman et al., *Upper Atmosphere and Space Programs in Canada: Special Study No. 1* (Ottawa: Science Secretariat, 1967) (Chapman Report). Some Canadian government reports refer to this program as the Canadian Rockets Development Program.

9 R.C. Fetherstonhaugh, *McGill University at War, 1914-1918 and 1939-1945* (Montreal and Kingston: McGill University Press, 1947), 321, 336-37. Early post-war missile studies are covered in Goodspeed, *A History of the Defence Research Board*, 127-33.

10 R.F. Wilkinson, "Rocket Research in Canada," *Canadian Aeronautical Journal* 5, 4 (April 1959): 138.

11 "(Confidential) Resume of Major DRB Activities Up to 1962, Scientific Program Rocket Propellant Research and Development," 3, RG 24, Accession (Acc.) 1983-84/167, file 7407 DRBS 173-2, pt. 1, LAC.

12 "Appendix A, DRB List of Technical Fields," July 1959, RG 24, Acc. 1983-84/167, file 73/778 3, LAC.

13 G.D. Watson to chief (sciences), (secret) memorandum reviewing progress of Defence Research Board activities, 24 June 1958, RG 24, Acc. 1983-84/167, file 7407 DRBS 173-1, LAC.

14 Proceedings of the second meeting of the Associate Committee on Space Research, Ottawa, "Annex C – Description of Black Brant I and II," by R. Blake, 8 April 1960, RG 25, vol. 7841, file 12798-2-40, pt. 1, LAC.

15 I.R. Cameron, "Manufacture and Testing of Black Brant Engines," *Canadian Aeronautical Journal* 7, 2 (February 1962): 61.

16 CARDE, "General Information," Technical Note (TN) 1421/61 (Unclassified), September 1961.

17 CARDE, "Summary of Performance of the 15KS25000 Rocket Engine Used in the First Sixteen Black Brant Vehicles," TN 1525/63, 1963, and CARDE, "The 15KS25000 Black Brant Engine Ground Operations and Handling Instructions," TN 1528/63, 1963.

18 CARDE, "Summary of Performance."

19 Cameron, "Manufacture and Testing," 65.
20 CARDE, "CARDE Black Brant I Vehicle Trials," Technical Memorandum 343/60, Valcartier, 1960.
21 S.F. Clark, chief of the general staff, Ottawa, "HQS 6001-Guided Missiles TD 8160 (CGS), Surface-to-Surface Missiles and Fort Churchill," 29 September 1959, (secret) memorandum, RG 25, file 1200, pt. 2 VII, LAC. For more details, see also A.M. Pennie, comp., *Defence Research Northern Laboratory, 1947-1965,* Report No. DR179 (Ottawa: Defence Research Board, 1966).
22 "Outer Space – Proposal for Possible Canadian Initiative, International Implications of a Canadian Space Research Program," 2, 27 August 1959, (confidential) memorandum prepared for the under-secretary of state for external affairs, RG 25, vol. 1, file 12798-4-40, LAC.
23 B. Claxton and R. Campney, *Operation Probe High, 20 July 1963* (Fort Churchill: Department of National Defence, 1963), 16-17.
24 Chapman Report, 22.
25 Cameron, "Manufacture and Testing," 66.
26 Ibid.
27 Ibid., 65.
28 The Black Brant was originally called the Snow Goose, a name that was later dropped from official documentation.
29 Department of External Affairs, NRC-DRB Permanent Joint Committee on Space Research, Proceedings of the first meeting of the Associate Committee on Space Research, Ottawa, 2, 2 October 1959, RG 25, vol. 7841, file 12798-2-40, pt. 1, LAC.
30 Albert W. Fia, "Canadian Sounding Rockets: Their History and Future Prospects," *Canadian Aeronautics and Space Institute Journal* 20, 8 (October 1974): 37-39.
31 Ibid. See also F. Jackson, "Development of the 23KS20000 Motor for the Black Brant IIB Vehicle," *Canadian Aeronautics and Space Journal* 11, 12 (December 1965): 377-83.
32 Donald C. Rose to E.W.R. Steacie, "Exhibit Q," 13 May 1959, written proposal for the formation of the ACSR, RG 25, vol. 7841, file 12798-2-40, pt. 1, LAC.
33 Ibid., 2.
34 Ibid., 5.
35 Ibid., 3.
36 Proceedings of the first meeting of the Executive Committee of the Associate Committee on Space Research, 1, 10 December 1959, RG 25, vol. 7841, file 12798-2-40, pt. 1, LAC, and "Outer Space – Proposal for Possible Canadian Initiative," 2.
37 Throughout this period, DND maintained a small cadre of military and civilian personnel in American missile-related programs and project offices.
38 "Outer Space – Proposal for Possible Canadian Initiative," 1.
39 Ibid., 2.
40 Proceedings of the second meeting of the Associate Committee on Space Research, 7, 8 April 1960, RG 25, vol. 7841, file 12798-2-40, pt. 1, LAC.
41 A fire, followed by politics, delayed the transition until 1 January 1966. The range continued to be funded and operated jointly by the United States and Canada until the late 1960s.
42 "Black Brant: Canadian Bristol Aerojet's Family of Sounding Rockets," *Flight International,* 7 January 1965, 14.
43 *Bristol Aerospace Limited,* 66.
44 Ibid., 15.
45 Ibid.

46 Peter Alway, *Rockets of the World,* 3rd ed. (Ann Arbor: Saturn Press, 1999), 343-46.
47 Fia, "Canadian Sounding Rockets," 398-99.
48 *Bristol Aerospace Limited,* 68.
49 "Black Brant: Canadian Bristol Aerojet's Family of Sounding Rockets," 17.
50 Alway, *Rockets of the World,* 343; see also Black Brant III Firings Table in "Black Brant: Canadian Bristol Aerojet's Family of Sounding Rockets," 17.
51 *Bristol Aerospace Limited,* 68, and Alway, *Rockets of the World,* 343.
52 *Bristol Aerospace Limited,* 70.
53 Ibid.
54 Ibid.
55 Ibid., 70-71.
56 Ibid., 71.
57 Some flights took place from the Northwest Territories, Peru, Brazil, Spain, Kauai Hawaii, and Greenland.
58 Ibid., 128.
59 Fia, "Canadian Sounding Rockets," 128, and *Bristol Aerospace Limited,* 71.
60 *Bristol Aerospace Limited,* 71. See also Chapman Report, 61-62.
61 Alway, *Rockets of the World,* 349-50.
62 Author e-mail interview with Lorne George Mason, Black Brant engineer (1963-65), 12 December 2000.
63 I.A. Stewart, "The Churchill Research Range and Canada's Satellite Program," *Canadian Aeronautics and Space Journal* 15 (October 1968): 311.
64 See L.H. Ohman, "A Satellite Launch Study Employing Black Brant VB Sounding Rocket Motors as Booster Building Blocks," *Canadian Aeronautics and Space Institute Journal* 13 (November 1967): 427.
65 Ibid., 427-29.
66 Stewart, "The Churchill Research Range," 311-13.
67 Chapman Report, 106.
68 Ibid., 105.
69 The Chapman Report provided an assessment of Black Brant activity but did not reveal any plans for creating an all-Canadian launch system.
70 H.R. Warren, "Design Considerations for CERES – A Satellite to Survey Canada's Natural Resources," *Canadian Aeronautics and Space Journal,* April 1968, 119-29.
71 Stewart, "The Churchill Research Range," 313, and Steven J. Isakowitz, *International Reference Guide to Space Launch Systems,* 2nd ed. (Washington, DC: American Institute of Aeronautics and Astronautics, 1991).
72 "Outer Space – Proposal for Possible Canadian Initiative, International Implications of a Canadian Space Research Program," 5, 27 August 1959.
73 Chapman Report, 102-4.
74 Richard K. Graf, "A Brief History of the HARP Project," *Encyclopedia Astronautica,* http://www.astronautix.com/articles/abroject.htm.
75 Ibid.
76 B. Sterling, "Think of the Prestige," from the *Magazine of Fantasy and Science Fiction,* September 1992, http://www-istp.gsfc.nasa.gov/stargaze/SGbull.htm.
77 F.W. Eyre, "The Development of Large Bore Gun Launched Rockets," *Canadian Aeronautics and Space Journal,* April 1966, 143-49.
78 J. Adams, *Bull's Eye: The Assassination and Life of Super Gun Inventor Gerald Bull* (New York: Times Books, 1992).

79 Ibid., 22-23.
80 Chapman Report, 62.
81 Ibid., 62-63, 102-5.

Chapter 3: Defence and Discovery

1 Untitled file, Record Group (RG) 25, file 12798-4-40, pt. 1, Library and Archives Canada (LAC), Ottawa.
2 "Canadians in the World – History of the Department of Foreign Affairs and International Trade, Dealing with Diefenbaker 1957-1963," Foreign Affairs and International Trade Canada, http://www.dfait-maeci.gc.ca/ciw-cdm/history-9-en.asp.
3 G.B. Doern, *Science and Politics in Canada* (Montreal and Kingston: McGill-Queen's University Press, 1972), 144.
4 Ibid.
5 Ibid., 4-5.
6 L.B. Pearson, *Mike: The Memoirs of the Rt. Hon. Lester B. Pearson,* 3 vols. (Toronto: University of Toronto Press, 1975).
7 Doern, *Science and Politics,* 145.
8 James R. Killian Jr., *Sputnik, Scientists, and Eisenhower* (Cambridge, MA: MIT Press, 1977), 2.
9 For Pearson's account of the period, see Pearson, *Mike,* vol. 2, *1948-1957;* see also John G. Diefenbaker, *One Canada: Memoirs of the Right Honourable John G. Diefenbaker,* 3 vols. (Toronto: Macmillan of Canada, 1976).
10 Canada, *House of Commons Debates* (1957-58).
11 "Some Factors Affecting Defence Research Policy – A Report to Board Members, October 1957," 1, 23 October 1957, (secret) memorandum, RG 24, file DRBS 173-1 (CDRB), LAC.
12 Ibid., 4.
13 D.V. LePan to Mr. Holmes, "International Control of Outer Space," 20 August 1958, memorandum, RG 25, file 12798-4-40, pt. 1, LAC.
14 Ibid.
15 A.G. Campbell, U.N. Division, to D.V. LePan, 12 September 1959, Department of External Affairs memorandum, RG 25, file 12798-4-40, pt. 1, LAC.
16 "Space Science and Space Technology – A Summary of Points Affecting Canada's Future Position," 17 December 1958, summary and paper, RG 25, DEA, box 112, file 4145-09-1, pt. 1, file DRBS 170-80/A16 (CDRB), LAC.
17 A.G. Campbell, U.N. Division, to Mr. Holmes, "Subject Outer Space," 3 March 1959, Department of External Affairs memorandum, RG 25, file 12798-4-40, pt. 1, LAC.
18 "(Confidential) Draft Memorandum to the Prime Minister on Outer Space – A Possible Canadian Initiative," 16 March 1959, RG 25, file 12798-4-40, pt. 1, LAC.
19 Ibid., 3.
20 "Proposal for Possible Canadian Initiative," 12 March 1959, (confidential) memorandum for the minister, RG 25, file 12798-4-40, pt. 1, LAC.
21 "(Confidential) Draft Memorandum to the Prime Minister on Outer Space – A Possible Canadian Initiative," 15 April 1959, with corrections, and (confidential) memorandum from the Office of the Secretary of State for External Affairs, signed by H.B. Robinson, 1 May 1959, RG 25, file 12798-4-40, pt. 1, LAC.
22 "International Cooperation in the Peaceful Uses of Outer Space," 1 June 1959, (confidential) letter to the president of the NRC and the chairman of the DRB, RG 24, file 12798-4-40, pt. 1, LAC.

23 Though no official date is given, the technical feasibility working group appears to have met between 10 June and 10 August 1959. Its final report was submitted on 27 August 1959.
24 "Under-Secretary's Discussion on Outer Space with the President of the National Research Council and the Chairman of the Defence Research Board," 19 June 1959, (confidential) memorandum, RG 25, file 12798-4-40, pt. 1, LAC.
25 Ibid., 2.
26 Ibid.
27 Ibid., 3.
28 Ibid.
29 "Outer Space – Proposal for Possible Canadian Initiative," 1959, RG 24, file 12798-4-40, pt. 1, LAC.
30 For details of Canada-US space cooperation agreements between 1945 and 1974, see A.B. Godefroy, "From Alliance to Dependence: Canadian-American Cooperation through Space, 1945-1999" (master's thesis, Royal Military College of Canada, Kingston, 1999); A.B. Godefroy, *Allies in Orbit: Canadian-American Defence Cooperation through Space* (Ottawa: Department of National Defence, Directorate of Space Development, 2000); and J.H. Chapman et al., *Upper Atmosphere and Space Programs in Canada: Special Study No. 1* (Ottawa: Science Secretariat, 1967) (Chapman Report).
31 For technical and legal details of these agreements, see the Chapman Report, 145-200.
32 Ibid., 2.3 Satellite Communications, 14-15.
33 Ibid., 201-13.
34 "Committees and Boards – Canadian Military Space Group, pt. 1, Proposed Agreement with France regarding Cooperation in Space Science," 30 November 1967, RG 24, Acc. 83-84/232 46, file 1150-110/M16, pt. 1, LAC, and "Proposed International Space Science Agreement – France," 5 December 1967, RG 24, Acc. 83-84/232 46, file 1150-110/M16, pt. 3, LAC.
35 For general notes on Canada and space law, see A. Beesley et al., "Canada's Contribution to Outer Space Law and Arms Control in Outer Space," in *Space Strategy: Three Dimensions*, ed. Brian MacDonald (Toronto: Canadian Institute of Strategic Studies, 1987), 94-110.
36 "Committees and Boards – Canadian Military Space Group, pt. 3, (Canadian Eyes Only) Memorandum from F.S.B. Thompson, Director Requirements Communications, to SA/CTS – Proposed International Space Science Agreement – France," 5 December 1967, RG 24, Acc. 83-84/232 46, file 1150-110/M16, pt. 1, LAC.
37 Canada, *Report of the Royal Commission on Government Organization* (Ottawa: Queen's Printer, 1963), 4:274.
38 Ibid.
39 Doern, *Science and Politics*, 211-24.
40 (Secret) recommendations for future action on the RCAF proposal for a Canadian space defence program, 20 July 1964, RG 24, S925-121-3, LAC.
41 Ibid.
42 Wing Commander K. Birchall (A/DAED) to D/AMTS, 28 September 1964, (secret) memorandum, RG 24, S925-121-3, LAC. The Tripartite Technology Cooperation Panel Sub-Group M meeting on military space was held 13-16 October 1964 in London, England.
43 Canada, Science Council of Canada, *A Space Program for Canada*, Report No. 1 (Ottawa: Queen's Printer, Science Council of Canada, 1967), 2.
44 Ibid., 3.

45 Ibid., 2.
46 Senate of Canada, *Committee on Science Policy,* No. 8. (Ottawa: Queen's Printer, 1968), 945-46. For related details, see also Canada, Science Council of Canada, *Towards a National Science Policy for Canada,* Report No. 4 (Ottawa: Queen's Printer, 1968), 10-20.
47 Doern, *Science and Politics,* 172.
48 Canada, Science Council of Canada, *Towards a National Science Policy,* 35.
49 R.L. Heilbroner, "The Impact of Technology: The Historical Debate," in *Automation and Technological Change,* ed. J.T. Dunlop (Englewood Cliffs, NJ: Prentice-Hall, 1962), 25.
50 Canada, Science Council of Canada, *A Space Program for Canada,* 7.

Chapter 4: Forging a Spacefaring Nation

1 The satellite's official name, as given in technical publications, was the S-27 Topside Sounder, or Beta Alpha One.
2 Because Alouette was the first satellite completely designed and built outside the USSR and the United States, this author argues that Canada should hold the title of the third country to place a man-made object into orbit.
3 For the technical history of the Alouette-ISIS program, see T.R. Hartz and I. Paghis, *Spacebound* (Ottawa: Department of Communications, Minister of Supply and Services Canada, 1982). For an analysis of the relationship between satellite communications and nation building in Canada, see P. Roper, "Seeking a Clearer Channel: Canadian Ventures in Satellite Technology and Nation Building, 1958-1972" (PhD thesis, University of Ottawa, 2003). Personal perspectives on space science and Alouette by defence scientists who served during this period may be drawn from the University of Western Ontario History of Space Science Workshop, 25 November 2002 (UWO Space Workshop), http://quark.physics.uwo.ca/~drm/history/space/space_history.html.
4 Joseph Schumpeter, *Business Cycles* (New York: McGraw-Hill, 1939); J. Schmookler, *Invention and Economic Growth* (Cambridge, MA: Harvard University Press, 1966).
5 John Vardalas, *The Computer Revolution in Canada: Building National Technological Competence* (Cambridge, MA: MIT Press, 2002), 276.
6 See J.J. Brown, *Ideas in Exile* (Toronto: McClelland and Stewart, 1967); Christian DeBresson, "Have Canadians Failed to Innovate? The Brown Thesis Revisited," *History of Science and Technology in Canada Bulletin* 6 (January 1982): 10-23; and Christian DeBresson and B. Murray, *The Supply and Use of Technological Innovation in Canada* (Ottawa: Cooperative Research Unit on Science and Technology, 1983).
7 LeRoy Nelms, "DRTE and Canada's Leap into Space: The Early Canadian Satellite Program," September 1991, Friends of CRC/Les amis du CRC (Communications Research Centre), http://www.friendsofcrc.ca/Articles/.
8 For references to US-Canada cooperation, see Homer E. Newell, *Beyond the Atmosphere: Early Years of Space Science,* NASA SP-4211 (Washington, DC: NASA, 1974), and J. Logsdon, ed., *Exploring the Unknown – Selected Documents in the History of the US Civil Space Program,* vol. 2, *External Relationships* (Washington, DC: NASA, 1996).
9 "Supplementary Materials: Alouette Satellite: Comments from Colin Hines," 22 September 1981, UWO Space Workshop. Hines was adding details to a letter from Forsyth to Scott on the origins of the Alouette concept.
10 Canada, Department of Communications, *Alouette 1: Canada's First Venture into Space* (Ottawa: Information Services, 1974), 6, and J.E. Jackson, R. Knecht, and S. Russell, "First Results in the NASA Topside Sounder Program," in United States, NASA, *Publications of Goddard Space Flight Center, 1959-1962,* vol. 2, *Space Technology* (Washington, DC: US Government Printing Office, 1966), 41.

11 Records infer that the meeting was held at Cornell University at Ithaca, New York, in October 1958. For details, see C.A. Franklin, "Alouette/ISIS: How It All Began" (address at IEEE International Milestone in Engineering Ceremony, Shirley's Bay, Ottawa, 13 May 1993).
12 Ibid.; see also Canada, Department of Communications, *Alouette 1,* 6.
13 Nelms, "DRTE and Canada's Leap into Space," 2.
14 Jackson, Knecht, and Russell, "First Results," 41-42.
15 Copies of these agreements may be found in J.H. Chapman et al., *Upper Atmosphere and Space Programs in Canada: Special Study No. 1* (Ottawa: Science Secretariat, 1967) (Chapman Report).
16 Jackson, Knecht, and Russell, "First Results," 41-42.
17 Ibid., 42-44.
18 L. Wallace, *Dreams, Hopes, Realities: NASA's Goddard Space Center, the First Forty Years,* NASA SP-4312 (Washington, DC: NASA History Office, 1999).
19 Stephen B. Johnson, *The Secret of Apollo: Systems Management in American and European Space Programs* (Baltimore: Johns Hopkins University Press, 2002), 2-4.
20 Technical specifications for Alouette may be found in the DRTE annual reports of 1962 through 1967; see also Canada, Department of Communications, *Alouette 1.* DRTE staff also published numerous technical papers in various scientific journals, and lists of these are available in the DRTE annual reports.
21 Canada, Defence Research Telecommunications Establishment, *Alouette Satellite 1962 Beta Alpha One* (Shirley's Bay: Defence Research Board, 1962).
22 J. Mar and H.R. Warren, "Structural and Thermal Design of the Topside Sounder Satellite," *Canadian Aeronautics and Space Journal,* September 1962, 163.
23 Canada, Department of Communications, *Alouette 1,* 13-15.
24 Mar and Warren, "Structural and Thermal Design," 163-64.
25 Ibid., 166-67.
26 Canada, Department of Communications, *Alouette 1,* 12-14.
27 Ibid., 12.
28 J. Mar, "Meteoroid Impact on the Topside Sounder Satellite," *Canadian Aeronautics and Space Journal,* November 1962, 237-40.
29 Canada, Department of Communications, *Alouette 1,* 14-20.
30 DRTE, "Annual Report," 1962. See also Mar and Warren, "Structural and Thermal Design," 164.
31 Mar and Warren, "Structural and Thermal Design," 165.
32 Quoted in Hartz and Paghis, *Spacebound,* 60-61.
33 L.J. Blumle, R.J. Fitzenreiter, and J.E. Jackson, "The National Aeronautics and Space Administration Topside Sounder Program," in United States, NASA, *Publications of Goddard Space Flight Center 1963,* vol. 2, *Space Technology* (Washington, DC: US Government Printing Office, 1965), 13.
34 "Supplementary Materials: Alouette Satellite: Comments from Colin Hines," UWO Space Workshop.
35 Nelms, "DRTE and Canada's Leap into Space," 23. See Chapter 7 for further details on the DRTE transfer to DOC.
36 Hartz and Paghis, *Spacebound,* 64.
37 Chapman Report, 11.
38 Ibid.
39 Hartz and Paghis, *Spacebound,* 65.
40 Chapman Report, 12.

41 The mission profile planned for an orbit apogee of three thousand kilometres. Hartz and Paghis, *Spacebound,* 65.
42 Ibid., 65-66.
43 T.G. Coughlin, "Alouette II – Canada's Second Satellite Probes the Mysteries of Outer Space," *Sentinel,* January-February 1966, 8-13.
44 Quoted in ibid., 8.
45 "Announcement of Opportunity for Participation in NASA 1971 Voyager Mission to Mars," 10 September 1965, DRTE paper, John H. Chapman Papers, MG 31, series J 43 14, file 14-2, Library and Archives Canada (LAC), Ottawa.
46 "(Confidential) Record of Cabinet Decision – International Satellites for Ionospheric Studies," 28 April 1964, Privy Council Office Papers, RG 2, LAC. An Order-in-Council (C. 1964-608) was passed accordingly.
47 Hartz and Paghis, *Spacebound,* 71.
48 C.D. Florida, *The ISIS Satellites,* DRTE Technical Note No. 619 (Ottawa: Department of Communications-Communications Research Center, 1969), 11.
49 Hartz and Paghis, *Spacebound,* 73.
50 Ibid., 74.
51 "Supplementary Materials: Alouette Satellite: Comments from Colin Hines," UWO Space Workshop. Hines described John Chapman as the man who "carried the can at DRB/HQ in getting the extra funding."
52 Florida, *The ISIS Satellites,* 15.
53 "Transcript: DRTE and UWO," 25 November 2002, UWO Space Workshop.
54 Ibid. Much of the discussion in this section of the UWO Space Workshop focused on the period between 1955 and 1965, during which Prime Minister Diefenbaker influenced many aspects of Canadian space research and development.

Chapter 5: The Militarization and Weaponization of Space in Canada

1 H. Bull, *The Control of the Arms Race: Disarmament and Arms Control in the Missile Age* (London: Weidenfeld and Nicolson, 1961), 1-6.
2 Paul Stares, *The Militarization of Space: U.S. Policy, 1945-1984* (Ithaca, NY: Cornell University Press, 1985), 17.
3 See Jane's Information Group, *Jane's Space Directory 2003-2004* (London: Jane's Information Group, 2003), 597. The Soviet anti-satellite program lasted from 1963 until 1982. For notes on this program, see Sven Grahn, "Simulated War in Space – Soviet ASAT Tests," http://www.svengrahn.pp.se/histind/ASAT/ASAT.htm.
4 For an overview of the main themes that have emerged in American historiography concerning relationships between the military, technological development, and economic and social change, see Merritt Roe Smith, ed., *Military Enterprise and Technological Change* (Cambridge, MA: MIT Press, 1987). See also John Vardalas, *The Computer Revolution in Canada: Building National Technological Competence* (Cambridge, MA: MIT Press, 2002), Introduction.
5 It is not my intent to detail the history of Canada and NORAD, as this is well addressed elsewhere. Among others, see J. Jockel, *No Boundaries Upstairs: Canada, the United States, and the Origins of North American Air Defence, 1945-1958* (Vancouver: UBC Press, 1987).
6 On the navy, see John Vardalas, "From DATAR to the FP-6000 Computer: Technological Change in a Canadian Industrial Context," *IEEE Annals of the History of Computing* 16, 2 (1994): 20-30, and John Vardalas, "The Navy's Pursuit of Self-Reliance in Digital Electronics," in Vardalas, *The Computer Revolution,* 45-78. On the army, see Sean M. Maloney, *An Identifiable Cult: The Evolution of Combat Development in the Canadian Army* (Kingston:

Department of National Defence, Directorate of Land Strategic Concepts, 1999). Pioneering efforts in the Canadian history of science and technology, these three works introduce often ignored aspects of post-war modernization of the Canadian military.

7 "(Secret) Defence Research Board Meeting Review of Progress," signed by DWR (director weapons research) G.D. Watson, 24 June 1958, RG 24, Acc. 83-84/DRB 173-1, Library and Archives Canada (LAC), Ottawa.

8 David Holloway, *Stalin and the Bomb: The Soviet Union and Atomic Energy, 1939-1956* (New Haven: Yale University Press, 1994), 314-17.

9 "Annex B to HQS 5025-C185, Coordination of the Joint Canadian/US Guided Missile Range," 11, October 1959, RG 24, vol. 25, file 1200, pt. 2, LAC.

10 L.C. Morrison, "South Atlantic Lookout: Canada Assists US Missile Research Program with Infrared Detection Team on Ascension Island," *Roundel,* April 1960, 2-3.

11 "The Cameras Saw Red," *Roundel,* March 1961, 13-14.

12 Quoted in ibid., 14.

13 DRB, "A Resume of Major Research Activities," 8 June 1962, RG 24, Acc. 83-84/167 7407 173-1, pt. 1, LAC.

14 "Defence Research Board Resume of Major Activities through June 1962," RG 24, Acc. 83-84/167 7407 173-1, pt. 1, LAC.

15 "Missile Monitors," *Roundel,* May 1962, 15. RCAF crews even conducted measurements on the American Atlas booster used to propel John Glenn on his way to becoming the first American to orbit the earth.

16 DRB, "A Resume of Major Research Activities," 4, 8 June 1962.

17 "Defence Research Board Resume of Major Activities through October 1962," RG 24, Acc. 83-84/167 7407 173-1, pt. 1, LAC.

18 Air Marshal Hugh Campbell, chief of air staff, to Air Marshal C.R. Slemon, deputy commander in chief of NORAD, 1, 28 March 1961, RG 24, 17829, file 840-105-001.8, LAC.

19 "(Secret) Development and Associated Research Policy Group Paper 12/65 – A Canadian Forces Space Development Program," 2, 12 May 1965, RG 24, 17973, file 925-121-3, L1150-4110/D8, LAC.

20 Campbell to Slemon, 2, 28 March 1961.

21 Ibid., 3.

22 Ibid., 4. Although they did not enter the US astronaut corps, RCAF officers did work at NASA. See the discussion of the Space Indoctrination Program later in this chapter.

23 J. Starnes, *Closely Guarded: A Life in Canadian Security and Intelligence* (Toronto: University of Toronto Press, 1998), 105-6.

24 Campbell to Slemon, 3, 28 March 1961.

25 For the only known published Canadian source on the SIP, see "Canadian Missile Men," *Sentinel,* September 1969, 32-33. Interviews with RCAF officers who were involved in the SIP have revealed that they were never informed that their "loan" to the United States fell under any organized plan, though official documentation and the above-mentioned article refer to the program.

26 "Reports and Returns – RCAF Personnel on Exchange Duties – USAF – HQS Air Force Space Systems Division, Los Angeles," c. 1963-65, RG 24, file 813-89/3-42, LAC.

27 "Exchange Officer Report, Flight Lieutenant A. Thoma, Headquarters Air Force Space Systems Division, Los Angeles," 15 March 1963, RG 24, file 813-89/3-42, LAC.

28 Dennis R. Jenkins, *Space Shuttle: The History of the National Space Transportation System* (Hong Kong: World Print, 2001), 30-31.

29 Author e-mail interview with Richard "Bud" White, 20 March 2003.

30 Ibid.

31 "RCAF Exchange Officer Report – F/L T.M. Harris and F/L J.H. Lathey," 28 February 1965, RG 24, file 813-89/3-42, LAC.

32 "(Confidential) Canadian Military Space Group First Report, Annex 1 – A Note on the SI," 31 January 1968, RG 24, Acc. 83-84/232 46, file 1150-110/M16, pt. 2, LAC. In 1967 seventeen officers had completed their SIP, and a further twelve were on exchange, as follows: boosters (four), environmental testing (three), flight dynamics (two), and one each with satellite communications, space surveillance, and Project START.

33 Ibid.

34 Curtis Peebles, *The Corona Project: America's First Spy Satellites* (Annapolis: Naval Institute Press, 1997), Appendices 1 and 2.

35 For an example, see B. Wooding and T.A. Spruston, "The Canadian Armed Forces and the Space Mission," *Canadian Defence Quarterly* 5, 2 (Winter 1975): 15-20.

36 Even four decades later, activities surrounding Canadian-American cooperation in the surveillance of space remain classified. Most of the material used for this section was derived via Access to Information. "Telecommunication Services – Data Processing – Ground Environment – Space Detection and Tracking System (SPADATS)," RG 24, vol. 17996, file 947-103-6, LAC.

37 "Annex A to 947-3-6 (DRDP), a Report on Exchange Duty with the USAF 496L System Program Office," 2, 26 November 1964, RG 24, vol. 17996, file 947-103-6, LAC.

38 Ibid.

39 K. Rodzinyak, "Like a Sapphire in the Sky: Canada's Surveillance of Space Project" (paper, War Studies Program, Royal Military College of Canada, Kingston, 2002).

40 Wooding and Spruston, "The Canadian Armed Forces," 17.

41 Vardalas, *The Computer Revolution*, 95-98.

42 Some unclassified Canadian assessments may be found in later defence literature. For an example, see S.L. Bennett, *Strategic Command, Control, and Communications: Capabilities, Doctrine, and Vulnerability* (Ottawa: Operational Research and Analysis Establishment, 1983). For technical information on Soviet launches during this period, see Jane's, *Space Directory*. Further descriptions may be found in Brian Harvey, *Russia in Space: The Failed Frontier?* (London: Springer-Praxis, 2001), 107-20.

43 For an unclassified Canadian overview of this study program, see K.J. Radford, "Studies of Orbital Rendezvous," *Canadian Aeronautics and Space Journal*, May 1962, 105-11. The program may have employed either the CRC102A computer or the recently developed DRTE computer.

44 "Advanced Engineering Space Systems Counter Satellite," RG 24, vol. 17973, file 925-121-3, LAC.

45 "Appendix A to S925-121-3 (Secret) Recommendations for Future Action on the RCAF Proposal for a Canadian Defence Space Program," 20 July 1964, RG 24, 17829, file 840-105-001.8, LAC.

46 Ibid., 6

47 Ibid., 6-7.

48 Ibid., 7-8.

49 These were Gemini VI and Gemini VIII through XII.

50 "(Secret) A Canadian Forces Space Development Program, Annex B," 4, 21 December 1964, supplement to RCAF proposal to the Development and Associated Research Policy Group (DARPG), RG 24, 17973, file 925-121-3, LAC.

51 Ibid., 3.

52 "Advanced Engineering Space Systems Counter-Satellite 1964-1965, (Secret) Appendix B – Summary of the Proposed Space Defence Program," 8, n.d., RG 24, 17973, file 925-121-3, LAC.

53 Ibid., 9. If the target was a manned spacecraft, an optical device was preferred to "facilitate the realization of rendezvous."
54 Ibid.
55 "Advanced Engineering Space Systems Counter-Satellite 1964-1965, (Secret) Memorandum S925-121-3 (DAED) Canadian Space Defence Program – Recommendations for Further Action," 1, 20 July 1964, RG 24, 17973, file 925-121-3, LAC.
56 Ibid., 2.
57 Ibid., 3.
58 Ibid., 4.
59 Ibid.
60 (Secret) minute sheet from Group Captain P.F. Peter (DAED) to A/AMTS, 21 July 1964, RG 24, S925-121-3, LAC.
61 Ibid.
62 Canada, *White Paper on Defence* (Ottawa: Queen's Printer, 1964); see also Douglas Bland, ed., *Canada's National Defence,* vol. 1, *Defence Policy* (Kingston: School of Policy Studies, Queen's University, 1997), 57-110.
63 (Secret) memorandum from Group Captain A.A. Buchanan, acting director general development, to CLED on DARPG submission, 15 April 1965, RG 24, S925-121-3, LAC.
64 C.L. Annis to Victor S.J. Millard, 30 September 1964, memorandum, RG 24, S925-121-3, LAC.
65 "(Secret) Temporary Docket 4248 – L1150-4110/DB DARPG Paper 12/65," 12 May 1965, RG 24, S925-121-3, LAC.
66 Ibid., 3-5.
67 Ibid., 4.
68 Ibid., 5.
69 (ScD/CLED/ED), J.C. Arnell to Mr. Wilkinson, Group Captain Peter, and Squadron Leader Flavin, 5 April 1965, memorandum, RG 24, S925-121-3 TD 4248, LAC. The memo called for a unanimous decision on future DND space priorities.
70 Lieutenant Colonel D.B.D. Warner (DACS) to A/CTS, "Advanced Engineering Space Systems Generally," 28 October 1965, memorandum, RG 24, file 925-121-2, LAC.
71 "Agenda of the 55th Meeting of the Defence Research Board," 12 March 1963, memorandum to the CNS, RG 24, Acc. 83-84/167 7407 173-1, pt. 2, LAC.
72 "Advanced Engineering Space Systems Generally (Confidential) S925-120-2 (DACS) Canadian Military Space Activities – Preliminary Report to the ScD/CTS on Related Items," 2-3, 28 October 1965, RG 24, file 925-121-2, LAC.
73 Ibid., 3.
74 Ibid., paragraph 7.

Chapter 6: The Demise of Canada's Cold War Space Programs

1 Merritt Roe Smith, ed., *Military Enterprise and Technological Change* (Cambridge, MA: MIT Press, 1987).
2 Douglas Bland, ed., *Canada's National Defence,* vol. 1, *Defence Policy* (Kingston: School of Policy Studies, Queen's University, 1997), 57.
3 Ibid., 57-58.
4 Ibid., 58.
5 "Report on Integration and Unification, 1964-68," 1968, R.L. Raymont Papers, file 759, Directorate of History and Heritage, Ottawa.
6 For more on this period, see G. Donaghy, *Tolerant Allies: Canada and the United States, 1963-1968* (Montreal and Kingston: McGill-Queen's University Press, 2002), 92-99.
7 Ibid., 100.

8 For a discussion of military assistance in Africa, see A.B. Godefroy, "The Canadian Armed Forces Advisory Training Team Tanzania, 1965-1970," *Canadian Military History* 11, 3 (Summer 2002): 31-47.
9 Ottawa to State Department, tel. no. 937, 21 January 1964, and State Department to Ottawa, tel. no. 614, 22 January 1964, Record Group (RG) 59, file Def 4, Canada-U.S., United States National Archives, Washington DC.
10 C.R. Slemon, "Meeting the Potential Soviet Threat from Space" (address to the USAF Association convention, Las Vegas, 1963), reprinted in *Roundel,* March 1963, 2-4.
11 "Review of CEPE Projects – Sample Technical Screening," 2, 2 March 1964, RG 24, 17973, file 925-121-3, Library and Archives Canada (LAC), Ottawa.
12 Ibid.
13 Ibid., 4.
14 "Advanced Engineering Space Systems Generally (Confidential) S925-120-2 (DACS) Canadian Military Space Activities – Preliminary Report to the ScD/CTS on Related Items," 3, 28 October 1965, RG 24, file 925-121-2, LAC.
15 Ibid.
16 "Review of CEPE Projects – Sample Technical Screening," 3, 2 March 1964.
17 Author e-mail interview with Squadron Leader Richard "Bud" White, RCAF, 18 March 2003.
18 "Advanced Engineering Space Systems, Generally, Canadian Military Space Activities – Preliminary Report to ScD/CTS on Related Items," 28 October 1965, RG 24, file 925, 121-2, LAC.
19 For a narrative of events, see B.S. Greaves, "Don't Write, Call Me via Satellite," *Sentinel,* March 1969, 6-9.
20 "Advanced Engineering Space Systems, Generally, Canadian Military Space Activities – Preliminary Report to ScD/CTS on Related Items," 5, 28 October 1965, RG 24, file 925, 121-2, LAC.
21 "(Confidential) Summary of Activities of the Defence Research Board," 2, 15 October 1964 to 15 February 1965, RG 24, Acc. 83-84/167 7407, 173-1, pt. 3, LAC.
22 "Annex II (Restricted), a Canadian Military Space Program – The Need for Action," RG 24, Acc. 83-83/232 46, file 1150-110/M16, pt. 1, LAC.
23 Cabinet met on 6 June and decided to pursue a program largely focused on domestic satellite communications.
24 "Annex II (Restricted), a Canadian Military Space Program," paragraph 1.
25 L.G.C. Lilley to ADM(L), 9 August 1967, (confidential) memorandum L-3390-1 (CTS), RG 24, Acc. 83-84/232 46, file 1150-110/M16, pt. 1, LAC.
26 "Committees and Boards – Canadian Military Space Group," 9 August 1967, memorandum L-3390-1 (SA/CTS) to DCENG, RG 24, Acc. 83-84/232 46, file 1150-110/M16, pt. 1, LAC.
27 Throughout its brief existence, the Canadian Military Space Group went by several monikers such as the Ad Hoc Group on Military Space and the Canadian Military Space Committee. See various memos and notes as follows: 11 August 1967 – Major A.S. Henry replies to CTS as VCDS is away at the time. Passed to Brigadier W.K. Lye (DGLF) on 23 August 1967 to resolve. The VCDS later recommends that Lieutenant Colonel L.H. Wylie (DLFORL) represents his office (file V-3390-1 TD 7222, 24 August 1967). 11 August 1967 – (Secret) memorandum to SA CTS from Major General D.A.G. Waldock (DCENG) nominating Group Captain P.F. Peter (DGAS) and Colonel M.H.F. Webber (DGCES) to represent him on the CMSG. 15 August 1967 – Memorandum from H.L. Meuser to CTS identifying F.S.B. Thompson, director of communications requirements, as his representative to the CMSG. Received by CTS on 21 August 1967 (L-3390-1, 15 August 1967).

23 August 1967 – Memorandum to DC OPS from H. Cayley for Commodore H.A. Porter (DG – Maritime Forces) regarding request for VCDS representation on the CMSG. Again, L.H. Wylie was recommended (V-3390-1 TD.7222A [DMFORS-5], 23 August 1967). 24 August 1967 – Memorandum for distribution to CTS, DC PLANS, DC OPS, and SA/VCDS from Commodore D.E. Samson (Sec VCDS) that Wylie has been nominated as the DC OPS member of the CMSG. DC PLANS representation was still being considered (V-3390-1 [Sec VCDS], 24 August 1967). 29 August 1967 – Memorandum from W. Petrie deputy chairman (scientific) DRB to CTS identifying Roy Doohoo as the DRB representative to the CMSG (file DRBC 170-80/S65 [DC(SC)], 29 August 1967), RG 24, Acc. 83-84/232 46, file 1150-110/M16, pt. 1, LAC.

28 L.G.C. Lilley to ADM(L), 9 August 1967 (confidential) memorandum L-3390-1 (CTS), RG 24, Acc. 83-84/232 46, file 1150-110/M16, pt. 1, LAC.

29 Memorandum L-3390-1 (SA/CTS) distributed for the first meeting of the CMSG, 8 September 1967, RG 24, Acc. 83-84/232 46, file 1150-110/M16, pt. 1, LAC.

30 Ibid.

31 Ibid.

32 Ibid., 2.

33 Fourth meeting of the Canadian Military Space Group, 2 November 1967, RG 24, Acc. 83-84/232 46, file 1150-110/M16, pt. 1, LAC.

34 Ibid.

35 Funding for a Canadian Space Agency was authorized nearly twenty years after this recommendation was presented to Cabinet.

36 Ibid., 3.

37 Ibid., 4.

38 "(Confidential) Canadian Military Space Group First Report," 31 January 1968, RG 24, Acc. 83-84/232 46, file 1150-110/M16, pt. 1 (1967-68), pt. 2. (1968-70), and pt. 3 (1967-68), LAC.

39 Ibid., pt. 2, 3.

40 Ibid., 7.

41 "Policy Guidance and CFP 200 – Implications of Military Space Studies," 18 March 1968, memorandum for distribution from Colonel J.C. Henry (DSFP-2) V 3120-16, RG 24, Acc. 83-84/232 46, file 1150-110/M16, pt. 3, LAC.

42 Ibid., 3, paragraph 6.

43 Ibid., 3.

44 Minutes of the thirteenth meeting of the CMSG held on 23 July 1969, RG 24, Acc. 83-84/232 46, file 1150-110/M16, pt. 2 (1968-70), LAC.

45 Memorandum from CTS to CMSG distribution list, 28 August 1969, RG 24, Acc. 83-84/232 46, file 1150-110/M16, pt. 2 (1968-70), LAC. The CTS special assistant was assigned to act as chair should the CMSG plan to meet in the future.

46 Literature exploring the historical relationship between Canada's military and technology during this period is not plentiful. Those sources that do provide some overview include K. Krause, *Arms and the State* (Cambridge: Cambridge University Press, 1992); R. Van Steenburg, "An Analysis of Canadian-American Defence Economic Cooperation," in *Canada's Industrial Base*, ed. D. Haglund (Kingston: Ronald Frye, 1988), 141-65; L. Aronsen, "Planning Canada's Economic Mobilization for War," *American Review of Canadian Studies* 15 (1985): 38-58; and L. Aronsen, "Canada's Postwar Rearmament: Another Look at American Theories of the Military-Industrial Complex," *Historical Papers/Communications Historiques* (1981): 175-96. Neither *Canadian Military History* nor *Scientia Canadensis* list a single published work on this subject.

47 John Vardalas, *The Computer Revolution in Canada: Building National Technological Competence* (Cambridge, MA: MIT Press, 2002), 277.
48 Assistant Deputy Minister L.M. Chesley to chairman Defence Research Board, 22 September 1955, memorandum, RG 24, Acc. 83-84/167, box 7368, file 170-80/E5, pt. 1, LAC. See also Vardalas, *The Computer Revolution,* 82.
49 Vardalas, *The Computer Revolution,* 85.
50 Ibid., 280.
51 "Policy Guidance and CFP 200 – Implications of Military Space Studies," 2, paragraph 3.d, 18 March 1968.
52 For studies on the history and evolution of American systems engineering and systems management, see Thomas P. Hughes, *Rescuing Prometheus* (New York: Pantheon Books, 1998), and Stephen B. Johnson, *The Secret of Apollo: Systems Management in American and European Space Programs* (Baltimore: Johns Hopkins University Press, 2002).
53 Author interview and exchange of notes with Joel J. Sokolsky, January-February 2004. See also his publication "Over There with Uncle Sam: Peacekeeping, the 'Trans-European Bargain,' and the Canadian Forces," in *What NATO for Canada?* Martello Papers 23, ed. David G. Haglund (Kingston: Queen's University Centre for International Relations, 2000), 15-36.
54 Ibid., interview and exchange of notes, January-February 2004.
55 David Haglund and Stéphane Roussel, "The Contradictions of Canadian Strategic Culture: 'Imperial' Commitments within a 'Democratic Alliance'" (paper presented at the biennial meeting of the American Association of Canadian Studies, Portland, OR, November 2002), 14.
56 Henry Kissinger, *White House Years* (Boston: Little Brown, 1979), 383.

Chapter 7: National Interests and a New Agenda for Outer Space

1 John Vardalas, *The Computer Revolution in Canada: Building National Technological Competence* (Cambridge, MA: MIT Press, 2002), 173-80.
2 Ibid., 173.
3 The American experience is examined in D. Mowrey and N. Rosenberg, *Paths of Innovation: Technological Change in 20th-Century America* (Cambridge: Cambridge University Press, 1998).
4 "(Confidential) Record of Cabinet Decision – Meeting of June 6th, 1967 – Government Policy on Satellites," 6 June 1967, prepared for CDS by R.J. Sutherland, Record Group (RG) 24, Acc. 83-84/232, 46, file 1150-110/M16, pt. 1, Library and Archives Canada (LAC), Ottawa.
5 Ibid., 1.
6 Canada, *White Paper on a Domestic Satellite Communications System for Canada* (Ottawa: Queen's Printer, 1968) (Drury White Paper).
7 Ibid., 42.
8 Ibid.
9 Ibid., 32.
10 Ibid., 46.
11 "Proposed White Paper on a Domestic Commercial Satellite Communication System for Canada," 3, 27 February 1968, Privy Council Office Papers (PCO), RG 2, series A-5-a, 6338, LAC.
12 Ibid., 1.
13 "(Confidential) Memorandum to Cabinet – Canadian Space Program: The Need for a Central Body," 9 July 1969, Cabinet Document 719-69, RG 25, DEA, accession box 112, file 4145-09-1, LAC (Need for Central Body).

14 J.H. Chapman et al., *Upper Atmosphere and Space Programs in Canada: Special Study No. 1* (Ottawa: Science Secretariat, 1967), 109-10 (Chapman Report). See also Canada, Science Council of Canada, *A Space Program for Canada,* Report No. 1 (Ottawa: Queen's Printer, Science Council of Canada, 1967), main conclusions.

15 Need for Central Body, 3-6.

16 "Confirmation of the Decisions of Cabinet Committees – An Interdepartmental Committee on Space," 4 December 1969, PCO, RG 2, series A-5-a, pt. 6340, LAC.

17 Need for Central Body; see also "Confirmation of the Decisions of Cabinet Committees – An Interdepartmental Committee on Space," 4 December 1969.

18 Canada, Interdepartmental Committee on Space, *Annual Report 1976* (Ottawa: Supply and Services, 1977), 5.

19 Chapman Report, 109-10; see also conclusion of Canada, Science Council of Canada, *A Space Program for Canada.*

20 J. Ghent, *Canadian Government Participation in International Science and Technology* (Ottawa: Norman Patterson School of International Affairs, 1979), 44-45.

21 Ibid., 46-47.

22 "Development of Policy and Coordination of Activities in Space Technology," 16 August 1972, (confidential) memorandum to Cabinet, Cabinet Doc. 924/72, RG 25, LAC.

23 United States, Space Task Group, *The Post-Apollo Space Program: Directions for the Future* (Washington, DC: Congressional Research Service, 1969), iii.

24 Quoted in Government of Canada, *A Foreign Policy for Canadians* (Ottawa: Department of External Affairs, 1970), 1.

25 C.M. Drury, "International Aspects of Possible Future Canadian Participation in Space Programs," *Canadian Aeronautics and Space Journal,* February 1971, 33, 35.

26 Ibid.

27 Ibid., 35.

28 For a detailed analysis of European space planning difficulties, see Stephen B. Johnson, *The Secret of Apollo: Systems Management in American and European Space Programs* (Baltimore: Johns Hopkins University Press, 2002). Of particular note is Chapter 6: "Organizing ELDO for Failure."

29 D. Simonelli, "Cooperation in Space," *European Community,* January-February 1978, 19; see also B. Valentine, "Obstacles to Space Cooperation: Europe and the Post-Apollo Experience," *Research Policy* 1 (1971-72): 21-29.

30 Ghent, *Canadian Government Participation,* 47.

31 Ibid., 18.

32 J. Hyndman, "National Interest and the New Look," *International Journal* 26 (Winter 1970-71): 5-6.

33 Nor was this transition a uniquely Canadian experience. The United States underwent a similar process between 1968 and 1974. See, for example, C. Twomey, "The McNamara Line and the Turning Point for Civilian Scientist-Advisors in American Defence Policy, 1966-1968," *Minerva* 37 (1999): 235-58, and C. Twomey, "The Vietnam War and the End of Civilian Scientist Advisors in Defence Policy," *Breakthroughs: Security Studies Program MIT* 9, 1 (Spring 2000): 12-20.

34 Twomey, "The McNamara Line," 238.

35 Y. Ezrahi, *The Descent of Icarus: Science and the Transformation of Contemporary Democracy* (Cambridge, MA: Harvard University Press, 1990), 15.

36 Hyndman, "National Interest," 5-6.

37 Ghent, *Canadian Government Participation,* 19.

38 Ibid.

39 Ibid.

40 "(Confidential) Memorandum to Cabinet – Development of Policy and Coordination of Activities in Space Technology," 16 August 1972.
41 Ibid., 3.
42 Ghent, *Canadian Government Participation,* 48.
43 For more details on this satellite-provider dispute, see David B. Dewitt and John J. Kirton, *Canada as a Principal Power* (Toronto: John Wiley and Sons, 1983), 330-45.

Conclusion

1 Canada, *Report of the Royal Commission on Government Organization* (Ottawa: Queen's Printer, 1963), 2:119.
2 Ibid., 4:208.

Note on Sources

1 H.D. Hellige, "Die gesellschaftlichen und historichen Grundlagen der Technikgestaltung als Gegenstand der Ingenieurausbildung," in *Lernen aus der Technikgeschichte,* ed. U. Troitzsch and W. König, cited in John Vardalas, *The Computer Revolution in Canada: Building National Technological Competence* (Cambridge, MA: MIT Press, 2002).
2 Alex Roland, with Philip Shiman, *Strategic Computing: DARPA and the Quest for Machine Intelligence, 1983-1993* (Cambridge, MA: MIT Press, 2002), 397.

Bibliography

Archives

Directorate of History and Heritage, Ottawa
R.L. Raymont Papers

Library and Archives Canada (LAC), Ottawa
John H. Chapman Papers (MG 31)
Privy Council Office Papers, Cabinet Conclusions, 1944-72 (RG 2)
Records of the Department of External Affairs (RG 25)
Records of the Department of National Defence (RG 24)
William Lyon Mackenzie King Diary

Government Documents

Canada

Canada in Space. Ottawa: Minister of Supply and Services, 1983.

CARDE (Canadian Armament Research and Development Establishment). "CARDE Black Brant I Vehicle Trials." Technical Memorandum 343/60. CARDE Memorandum, Valcartier, April 1960.

–. "The 15KS25000 Black Brant Engine Ground Operations and Handling Instructions. Technical Note 1528/63." CARDE Technical Note, Valcartier, 1963.

–. "General Information. Technical Note 1421/61." CARDE Technical Note, Valcartier, September 1961.

–. "Summary of Performance of the 15KS25000 Rocket Engine Used in the First Sixteen Black Brant Vehicles. Technical Note 1525/63." CARDE Technical Note, Valcartier, 1963.

Defence Research Telecommunications Establishment. *Alouette Satellite 1962 Beta Alpha One.* Shirley's Bay: Defence Research Board, 1962.

–. "Annual Report." 1962, 1963, 1964, 1965, 1966, 1967.

Department of Communications. *Alouette 1: Canada's First Venture into Space.* Ottawa: Information Services, 1974.

Department of Reconstruction and Supply. *Research and Scientific Activity, Canadian Federal Expenditures, 1938-1946.* Ottawa: King's Printer, 1947.

Interdepartmental Committee on Space. *Annual Report 1976.* Ottawa: Supply and Services, 1977.

National Research Council. *Annual Report.* Ottawa: King's Printer, 1943-44, 1944-45, 1945-46.

Royal Canadian Air Force. *Staff College and Air Force College Journal.* 1956, 1957, 1958, 1959, 1960, 1961, 1962, 1963, 1964.

Science Council of Canada. *A Space Program for Canada.* Report No. 1. Ottawa: Queen's Printer, Science Council of Canada, 1967.

United States

NASA. *Ariel I: The First International Satellite.* NASA SP-43. Washington, DC: NASA, 1963.

–. *Publications of Goddard Space Flight Center 1959-1962.* Washington, DC: US Government Printing Office, 1966.

–. "World's First International Satellite." News release, April 1962.

Space Task Group. *The Post-Apollo Space Program: Directions for the Future.* Washington, DC: Congressional Research Service, 1969.

United States Space Command. Satellite Catalog. 1958, 1959, 1960, 1961, 1962.

Secondary Sources

Adams, J. *Bull's Eye: The Assassination and Life of Super Gun Inventor Gerald Bull.* New York: Times Books, 1992.

Alway, Peter. *Rockets of the World,* 3rd ed. Ann Arbor: Saturn Press, 1999.

Andrew, Christopher, and Vasili Mitrokhin. *The Mitrokhin Archive: The KGB in Europe and the West.* London: Penguin Press, 1999 and 2000.

Armstrong, A.W. "The RCAF Approach to the Evaluation of a Supersonic Interceptor Armed with Air-to-Air Guided Missiles." *Canadian Aeronautical Journal* 4,6 (June 1958): 194-99.

Aronsen, L. "Canada's Postwar Rearmament: Another Look at American Theories of the Military-Industrial Complex." *Historical Papers/Communications Historiques* (1981): 175-96.

–. "Planning Canada's Economic Mobilization for War." *American Review of Canadian Studies* 15 (1985): 38-58.

Avery, Donald H. *The Science of War: Canadian Scientists and Allied Military Technology during the Second World War.* Toronto: University of Toronto Press, 1998.

Babbit, J.D. *Science in Canada: Selections from the Speeches of E.W.R. Steacie.* Toronto: University of Toronto Press, 1965.

Beesley, A., and D.W. Sproule. "Canada's Contribution to Outer Space Law and Arms Control in Outer Space." In *Space Strategy: Three Dimensions,* ed. Brian MacDonald, 94-110. Toronto: Canadian Institute of Strategic Studies, 1987.

Bennett, S.L. *Strategic Command, Control, and Communications: Capabilities, Doctrine, and Vulnerability.* Ottawa: Operational Research and Analysis Establishment, 1983.

Berger, Carl. *The Writing of Canadian History: Aspects of English-Canadian Historical Writing since 1900.* Toronto: University of Toronto Press, 1993.

"Black Brant: Canadian Bristol Aerojet's Family of Sounding Rockets." *Flight International,* 7 January 1965, 14-17.

Bland, Douglas, ed. *Canada's National Defence.* Vol. 1, *Defence Policy.* Kingston: School of Policy Studies, Queen's University, 1997.

Blumle, L.J., R.J. Fitzenreiter, and J.E. Jackson. "The National Aeronautics and Space Administration Topside Sounder Program." In NASA, *Publications of Goddard Space Flight Center 1963.* Vol. 2, *Space Technology,* 112-35. Washington, DC: US Government Printing Office, 1965.

Bristol Aerospace Limited: 50 Years of Technology, 1930-1980. Vol. 2, *The Second Quarter Century.* Winnipeg: Bristol Aerospace, 1985.

Brown, J.J. *Ideas in Exile: A History of Canadian Invention.* Toronto: McClelland and Stewart, 1967.

Bryden, John. *Deadly Allies: Canada's Secret War, 1937-1947.* Toronto: McClelland and Stewart, 1989.

Bulkeley, Rip. *The Sputniks Crisis and Early United States Space Policy: A Critique of the Historiography of Space.* Bloomington: Indiana University Press, 1991.

Bull, H. *The Control of the Arms Race: Disarmament and Arms Control in the Missile Age.* London: Weidenfeld and Nicolson, 1961.

Cameron, I.R. "Manufacture and Testing of Black Brant Engines." *Canadian Aeronautical Journal* 7, 2 (February 1962): 61-66.

Chaikin, Andrew. *A Man on the Moon.* New York: Penguin Books, 1994.

Chapman, I., and M. Gibbons. "Innovation and the Senate Report on Science Policy." *Journal of Canadian Studies* 13,1 (Spring 1978): 30-37.

Chapman, J.H., P.A. Forsyth, P.A. Lapp, and G.N. Patterson. *Upper Atmosphere and Space Programs in Canada: Special Study No. 1.* Ottawa: Science Secretariat, 1967.

Chapman, Sidney. *IGY: Year of Discovery.* Ann Arbor: University of Michigan Press, 1959.

Chapnick, A. "The Canadian Middle Power Myth." *International Journal* 55, 2 (Spring 2000): 188-206.

–. "Principle for Profit: The Functional Principle and the Development of Canadian Foreign Policy, 1943-1947." *Journal of Canadian Studies* 37, 2 (Summer 2002): 68-85.

Chinnick, R.F. "Upper Atmosphere and Space Research in Canada." *Canadian Aeronautics and Space Journal* 8,10 (October 1962): 198-213.

Clauser, F.H. "Flight beyond the Earth." *Roundel,* November 1948, 10-14.

Claxton, B., and R. Campney. *Operation Probe High, 20 July 1963.* Fort Churchill: Department of National Defence, 1963.

Coughlin, T.G. "Alouette II – Canada's Second Satellite Probes the Mysteries of Outer Space." *Sentinel,* January-February 1966, 8-13.

DeBresson, Christian. "Have Canadians Failed to Innovate? The Brown Thesis Revisited." *History of Science and Technology in Canada Bulletin* 6 (January 1982): 10-23.

DeBresson, Christian, and B. Murray. *The Supply and Use of Technological Innovation in Canada.* Ottawa: Cooperative Research Unit on Science and Technology, 1983.

"Defence Research Board Tests: Rockets Feature 1958 Experiments." *Canadian Army Journal* 13, 1 (January 1959): 17-24.

Demorest, C.P. "Black Brant IIA Preparation and Launching Data." CARDE Technical Note 1728/66. Valcartier, August 1966.

Dewitt, David B., and John J. Kirton. *Canada as a Principal Power.* Toronto: John Wiley and Sons, 1983.

Diefenbaker, John G. *One Canada: Memoirs of the Right Honourable John G. Diefenbaker.* 3 vols. Toronto: Macmillan of Canada, 1976.

Doern, G.B. *Science and Politics in Canada.* Montreal and Kingston: McGill-Queen's University Press, 1972.

Donaghy, G. *Tolerant Allies: Canada and the United States, 1963-1968.* Montreal and Kingston: McGill-Queen's University Press, 2002.

"DRB Pioneers Missile Study." *Canadian Army Journal* 15,3 (Summer 1961): 54.

Drury, C.M. "International Aspects of Possible Future Canadian Participation in Space Programs." *Canadian Aeronautics and Space Journal,* February 1971, 33-35.

Eggleston, W. *National Research in Canada: The NRC, 1916-1966.* Toronto: Clarke Irwin, 1978.

Eyre, F.W. "The Development of Large Bore Gun Launched Rockets." *Canadian Aeronautics and Space Journal,* April 1966, 143-49.

Ezrahi, Y. *The Descent of Icarus: Science and the Transformation of Contemporary Democracy.* Cambridge, MA: Harvard University Press, 1990.

Feklisov, Alexsander. *The Man behind the Rosenbergs*. New York: Enigma Books, 2001.
Fetherstonhaugh, R.C. *McGill University at War, 1914-1918 and 1939-1945*. Montreal and Kingston: McGill University Press, 1947.
Fia, Albert W. "Canadian Sounding Rockets: Their History and Future Prospects." *Canadian Aeronautics and Space Institute Journal* 20, 8 (October 1974): 32-42.
Florida, C.D. "The ISIS Satellites." DRTE Technical Note No. 619. Ottawa, Department of Communications-Communications Research Centre, 1969.
Fordyce, A.M. "How It All Started: The Goforth Paper." *Canadian Defence Quarterly* 1, 4 (Spring 1972): 15-16.
Forman, Paul. "Behind Quantum Electronics: National Security as a Basis for Physical Research in the United States, 1940-1960." *Historical Studies in the Physical and Biological Sciences* 18, 1 (1987): 149-229.
Franklin, C.A. "Alouette/ISIS: How It All Began." Address at IEEE International Milestone in Engineering Ceremony, Shirley's Bay, Ottawa, 13 May 1993.
Freedman, Lawrence. *The Evolution of Nuclear Strategy*. New York: St. Martin's Press, 1989.
Gainor, Chris. *Arrows to the Moon: AVRO's Engineers and the Space Race*. Burlington: Apogee Books, 2001.
–. "Owen E. Maynard: Lunar Module Pioneer from Canada." *Quest: The History of Space Flight Quarterly* 9, 3 (2002): 22-28.
Gelly, A., and H.P. Tardiff. *Defence Research Establishment Valcartier, 1945-1995*. Ottawa: Minister of Public Works and Government Services Canada, 1995.
Ghent, J. *Canadian Government Participation in International Science and Technology*. Ottawa: Norman Patterson School of International Affairs, 1979.
Gingras, Yves. *Physics and the Rise of Scientific Research in Canada*. Montreal and Kingston: McGill-Queen's University Press, 1991.
Godefroy, A.B. *Allies in Orbit: Canadian-American Defence Cooperation through Space*. Ottawa: Department of National Defence, Directorate of Space Development, 2000.
–. "Canada's Early Space Policy Development, 1958-1974." *Space Policy Journal* 19 (2003): 137-41.
–. "From Alliance to Dependence: Canadian-American Cooperation through Space, 1945-1999." Master's thesis, Royal Military College of Canada, Kingston, 1999.
Goodspeed, D.J. *A History of the Defence Research Board of Canada*. Ottawa: Queen's Printer, 1958.
Graf, Richard K. "A Brief History of the HARP Project." *Encyclopedia Astronautica*. http://www.astronautix.com/articles/abrobject.htm.
Grahn, Sven. *A Man of Influence: Norman A. Robertson and Canadian Statecraft, 1929-1968*. Toronto: Deneau, 1981.
–. "Simulated War in Space – Soviet ASAT Tests." http://www.svengrahn.pp.se/histind/ASAT/ASAT.htm.
Granatstein, J.L., and D. Stafford. *Spy Wars: Espionage and Canada from Gouzenko to Glasnost*. Toronto: Key Porter, 1990.
Grant, Dale. *Wilderness of Mirrors: The Life of Gerald Bull*. Scarborough: Prentice-Hall Canada, 1991.
Green, J.J. "Science and Defence." *Canadian Aeronautical Journal*, April 1955, 3-6.
Haglund, David, and Stéphane Roussel. "The Contradictions of Canadian Strategic Culture: 'Imperial' Commitments within a 'Democratic Alliance.'" Paper presented at the biennial meeting of the American Association of Canadian Studies, Portland, OR, November 2002.

Hart, John Edward. "Canada and North American Defence, 1940-1965." Master's thesis, University of Oregon, 1967.

Hartz, T.R., and I. Paghis. *Spacebound.* Ottawa: Department of Communications, Minister of Supply and Services Canada, 1982.

Harvey, Brian. *Russia in Space: The Failed Frontier?* London: Springer-Praxis, 2001.

Haynes, John Earl, and Harvey Klehr. *Venona: Decoding Soviet Espionage in America.* New Haven: Yale University Press, 1999.

Heilbroner, R.L. "The Impact of Technology: The Historical Debate." In *Automation and Technological Change,* ed. J.T. Dunlop, 31-48. Englewood Cliffs, NJ: Prentice-Hall, 1962.

Hellyer, P. *Damn the Torpedoes: My Fight to Unify Canada's Armed Forces.* Toronto: McClelland and Stewart, 1990.

Heppenheimer, T.A. *Countdown: A History of Spaceflight.* New York: J. Wiley and Sons, 1997.

Hill, C.N. *A Vertical Empire: The History of the UK Rocket and Space Programme, 1950-1971.* London: Imperial College Press, 2001.

Holloway, David. *Stalin and the Bomb: The Soviet Union and Atomic Energy, 1939-1956.* New Haven: Yale University Press, 1994.

Hughes, Thomas P. *Rescuing Prometheus.* New York: Pantheon Books, 1998.

Hyndman, J. "National Interest and the New Look." *International Journal* 26 (Winter 1970-71): 5-18.

Isakowitz, Steven J. *International Reference Guide to Space Launch Systems.* 2nd ed. Washington, DC: American Institute of Aeronautics and Astronautics, 1991.

Jackson, F. "Development of the 23KS20000 Motor for the Black Brant IIB Vehicle." *Canadian Aeronautics and Space Journal* 11, 12 (December 1965): 377-83.

Jackson, J.E., R. Knecht, and S. Russell. "First Results in the NASA Topside Sounder Program." In NASA, *Publications of Goddard Space Flight Center, 1959-1962.* Vol. 2, *Space Technology,* 40-50. Washington, DC: US Government Printing Office, 1966.

Jane's Information Group. *Jane's Space Directory 2003-2004.* London: Jane's Information Group, 2003.

Jarrell, Richard A. *The Cold Light of Dawn: A History of Canadian Astronomy.* Toronto: University of Toronto Press, 1988.

Jarrell, Richard A., and Yves Gingras, eds. *Building Canadian Science: The Role of the National Research Council.* Ottawa: Canadian Science and Technology Historical Association, 1991.

Jelly, Doris. *Canada: 25 Years in Space.* Ottawa: National Museum of Science and Technology, 1988.

Jockel, J. *No Boundaries Upstairs: Canada, the United States, and the Origins of North American Air Defence, 1945-1958.* Vancouver: UBC Press, 1987.

Johnson, Stephen B. *The Secret of Apollo: Systems Management in American and European Space Programs.* Baltimore: Johns Hopkins University Press, 2002.

Jones, Greta. *Science, Politics, and the Cold War.* New York: Routledge, 1988.

Kevles, D. "Cold War and Hot Physics: Science, Security and the American State, 1945-1956." *Historical Studies in the Physical and Biological Sciences* 20 (1980): 239-64.

Killian, James R., Jr. *Sputnik, Scientists, and Eisenhower: A Memoir of the First Special Assistant to the President for Science and Technology.* Cambridge, MA: MIT Press, 1977.

King, A., et al. *Reviews of National Science Policy: Canada.* Paris: Organisation for Economic Co-operation and Development, 1969.

Kirton, J., ed. *Canada, the United States, and Space.* Toronto: Canadian Institute of International Affairs, 1986.

Kissinger, Henry. *White House Years.* Boston: Little Brown, 1979.

Koerner, Stephen. "Canada and the Post-War Reparations Programme." Paper presented at the Canadian Historical Association annual meeting, Edmonton, 27 May 2000, and the Canadian Society for the History and Philosophy of Science conference, 2001.

Krause, K. *Arms and the State.* Cambridge: Cambridge University Press, 1992.

Law, C.E. "Omond McKillop Solandt: Operational Research Pioneer." *INFOR* 31, 2 (May 1993): 9-14.

Law, C.E., G.R. Lindsey, and D.M. Grenville, eds. *Perspectives in Science and Technology: The Legacy of Omond Solandt.* Kingston: Queen's Quarterly, 1994.

Leslie, Stuart W. *The Cold War and American Science: The Military-Industrial-Academic Complex at MIT and Stanford.* New York: Columbia University Press, 1993.

LeSueur, Ernest A. "Rocketeers." *Canadian Defence Quarterly* 9, 3 (April 1932): 374-82.

Lindsey, George R., ed. *No Day Long Enough: Canadian Science in World War II.* Toronto: Canadian Institute of Strategic Studies, 1997.

Lithwick, N.H. *Canada's Science Policy and the Economy.* Toronto: Methuen, 1969.

Logsdon, J., ed. *Exploring the Unknown – Selected Documents in the History of the US Civil Space Program.* Vol. 2, *External Relationships.* Washington, DC: NASA, 1996.

Lowther, William. *Arms and the Man: Dr. Gerald Bull, Iraq, and the Supergun.* Toronto: Doubleday Canada, 1991.

Maloney, Sean M. *An Identifiable Cult: The Evolution of Combat Development in the Canadian Army.* Kingston: Department of National Defence, Directorate of Land Strategic Concepts, 1999.

–. "Learning to Love the Bomb: Canada's Cold War Strategy and Nuclear Weapons, 1951-1968." Manuscript based on PhD diss., Temple University, PA, 1996.

Mar, J. "Meteoroid Impact on the Topside Sounder Satellite." *Canadian Aeronautics and Space Journal* (November 1962): 237-40.

Mar, J., and H.R. Warren. "Structural and Thermal Design of the Topside Sounder Satellite." *Canadian Aeronautics and Space Journal,* September 1962, 161-69.

McDougall, W. *The Heavens and Earth: A Political History of the Space Age.* New York: Basic Books, 1985.

McLin, Jon B. *Canada's Changing Defence Policy, 1957-1963: The Problems of a Middle Power in Alliance.* Baltimore: Johns Hopkins Press, 1967.

Morrison, L.C. "South Atlantic Lookout: Canada Assists US Missile Research Program with Infrared Detection Team on Ascension Island." *Roundel,* April 1960, 2-3.

Mowrey, D., and N. Rosenberg. *Paths of Innovation: Technological Change in 20th-Century America.* Cambridge: Cambridge University Press, 1998.

National Science Foundation. *Science, Technology, and Democracy in the Cold War and After: A Strategic Plan for Research in Science and Technology Studies.* New Mexico: Santa Fe Institute Workshop, 1994.

Needell, Allan A. *Science, Cold War and the American State: Lloyd V. Berkner and the Balance of Professional Ideals.* Washington, DC: National Air and Space Museum, 2000.

Nelms, LeRoy. "DRTE and Canada's Leap into Space: The Early Canadian Satellite Program." September 1991. Friends of CRC/Les amis du CRC (Communications Research Centre). http://www.friendsofcrc.ca/Articles/.

Newell, Homer E. *Beyond the Atmosphere: Early Years of Space Science.* NASA SP-4211. Washington, DC: NASA, 1974.

Nicolet, Marcel. "Historical Aspects of the IGY." *Eos Transactions* 64, 19 (10 May 1983): 7-14.

Ohman, L.H. "A Satellite Launch Study Employing Black Brant VB Sounding Rocket Motors as Booster Building Blocks." *Canadian Aeronautics and Space Institute Journal* 13 (November 1967): 427.

Owram, Douglas. *Canadian History: A Reader's Guide.* Toronto: University of Toronto Press, 1994.

Pearson, L.B. *Mike: The Memoirs of the Rt. Hon. Lester B. Pearson.* 3 vols. Toronto: University of Toronto Press, 1975.

Peebles, Curtis. *The Corona Project: America's First Spy Satellites.* Annapolis: Naval Institute Press, 1997.

Penick, J.L., Jr., C.W. Pursell Jr., M.B. Sherwood, and Donald C. Swain, eds. *The Politics of American Science: 1939 to the Present.* Chicago: Rand McNally, 1965.

Pennie, A.M., comp. *Defence Research Northern Laboratory, 1947-1965.* Report No. DR179. Ottawa: Defence Research Board, 1966.

–. "A Quarter Century of Achievement." *Canadian Defence Quarterly* 1, 4 (Spring 1972): 6-14.

Radford, K.J. "Studies of Orbital Rendezvous." *Canadian Aeronautics and Space Journal*, May 1962, 105-11.

Radosh, Ronald, and Joyce Milton. *The Rosenberg File.* New Haven: Yale University Press, 1997.

Richmond, R.D., and J.F. Perrier. "The Technology of Guided Missiles and Its Effect on Industry." *Canadian Aeronautical Journal* 3,4 (April 1957): 113-20.

Richter, Andrew. *Avoiding Armageddon: Canadian Military Strategy and Nuclear Weapons, 1950-1963.* Vancouver: UBC Press, 2002.

–. *The Evolution of Strategic Thinking at the Canadian Department of National Defence, 1950-1960.* Occasional Paper 38. Toronto: Centre for International and Strategic Studies, York University, 1996.

Roberts, A.K. "Development of the 9KS11000 Black Brant III Rocket Engine." *Canadian Aeronautics and Space Journal* (June 1962): 137-43.

Rodzinyak, K. "Like a Sapphire in the Sky: Canada's Surveillance of Space Project." Paper, War Studies Program, Royal Military College of Canada, Kingston, 2002.

Roland, Alex, with Philip Shiman. *Strategic Computing: DARPA and the Quest for Machine Intelligence, 1983-1993.* Cambridge, MA: MIT Press, 2002.

Roper, P. "Seeking a Clearer Channel: Canadian Ventures in Satellite Technology and Nation Building, 1958-1972." PhD thesis, University of Ottawa, 2003.

Rose, D.C. "Space Research in Canada." *Canadian Aeronautics and Space Journal* 10, 4 (April 1964): 89-92.

Sawatsky, J. *Men in the Shadows: The RCMP Security Service.* Toronto: Doubleday Canada, 1980.

Schmookler, J. *Invention and Economic Growth.* Cambridge, MA: Harvard University Press, 1966.

Schumpeter, Joseph. *Business Cycles.* New York: McGraw-Hill, 1939.

Siddiqi, Asif A. *Challenge to Apollo: The Soviet Union and the Space Race, 1945-1974.* Washington, DC: NASA, 2000.

Simonelli, D. "Cooperation in Space." *European Community,* January-February 1978, 9-12.

Slemon, C.R. "Meeting the Potential Soviet Threat from Space." Address to the USAF Association convention, Las Vegas, 1963. Reprinted in *Roundel* (March 1963): 2-4.

Smith, Merritt Roe, ed. *Military Enterprise and Technological Change.* Cambridge, MA: MIT Press, 1987.

Snow, C.P. *Science and Government.* Cambridge, MA: Harvard University Press, 1961.

Sokolsky, Joel. "Over There with Uncle Sam: Peacekeeping, the 'Trans-European Bargain,' and the Canadian Forces." In *What NATO for Canada?* Martello Papers 23, ed. David G. Haglund, 15-36. Kingston: Queen's University Centre for International Relations, 2000.

Solandt, O.M. "The Defence Research Board's Untimely End: What It Means for Military Science." *Science Forum* 8 (October 1975): 19-21.

–. "Science and Industry in Canada." *Canadian Aeronautics and Space Journal* 8, 9 (September 1962): 157-59.

Spencer, Robert A. *Canada in World Affairs, 1946-1949.* Toronto: Macmillan Books, 1959.

Starnes, J. *Closely Guarded: A Life in Canadian Security and Intelligence.* Toronto: University of Toronto Press, 1998.

Sterling, B. "Think of the Prestige." From the *Magazine of Fantasy and Science Fiction,* September 1992. http://www-istp.gsfc.nasa.gov/stargaze/SGbull.htm.

Stewart, Greig. *Shutting Down the National Dream: A.V. Roe and the Tragedy of the AVRO Arrow.* Toronto: McGraw-Hill Ryerson, 1997.

Stewart, I.A. "The Churchill Research Range and Canada's Satellite Program." *Canadian Aeronautics and Space Journal* 15 (October 1968): 307-13.

Story, Donald C., and Russell Isinger. "The Origins of the Cancellation of Canada's Avro CF-105 Arrow Fighter Program: A Failure of Strategy." *Journal of Strategic Studies* 30, 6 (December 2007): 1025-50.

Sutherland, R.J. "Canada's Long Term Strategic Situation." *International Journal* 17, 3 (Summer 1962): 180-99.

Thistle, Mel. *The Inner Ring: The Early History of the National Research Council of Canada.* Toronto: University of Toronto Press, 1966.

Twomey, C. "The McNamara Line and the Turning Point for Civilian Scientist-Advisors in American Defence Policy, 1966-1968." *Minerva* 37 (1999): 235-58.

–. "The Vietnam War and the End of Civilian Scientist Advisors in Defence Policy." *Breakthroughs: Security Studies Program MIT* 9, 1 (Spring 2000): 12-20.

Ulman, William A. "Russian Planes Are Raiding Canadian Skies: Special Report on US Air Defenses." *Collier's,* 16 October 1953, 33-45.

University of Western Ontario. History of Space Science Workshop. 25 November 2002. http://quark.physics.uwo.ca/~drm/history/space/space_history.html.

Valentine, B. "Obstacles to Space Cooperation: Europe and the Post-Apollo Experience." *Research Policy* 1 (1971-72): 21-29.

Van Steenburg, R. "An Analysis of Canadian-American Defence Economic Cooperation." In *Canada's Industrial Base,* ed. D. Haglund and J.M. Treddenick, 141-65. Kingston: Ronald Frye, 1988.

Vardalas, John. *The Computer Revolution in Canada: Building National Technological Competence.* Cambridge, MA: MIT Press, 2002.

–. "From DATAR to the FP-6000 Computer: Technological Change in a Canadian Industrial Context." *IEEE Annals of the History of Computing* 16, 2 (1994): 20-30.

Wallace, L. *Dreams, Hopes, Realities: NASA's Goddard Space Center, the First Forty Years.* NASA SP-4312. Washington, DC: NASA History Office, 1999.

Warren, H.R. "Design Considerations for CERES – A Satellite to Survey Canada's Natural Resources." *Canadian Aeronautics and Space Journal,* April 1968, 119-29.

Watson, G.D. "The Scientific Exploration of Space." *Canadian Aeronautical Journal* 6, 3 (March 1960): 81-89.

Whitaker, Reg, and Gary Marcuse. *Cold War Canada: The Making of a National Insecurity State, 1945-1957.* Toronto: University of Toronto Press, 1994.
Wilkinson, R.F. "Rocket Research in Canada." *Canadian Aeronautical Journal* 5, 4 (April 1959): 138-42.
Wooding, B., and T.A. Spruston. "The Canadian Armed Forces and the Space Mission." *Canadian Defence Quarterly* 5, 2 (Winter 1975): 15-20.
Zimmerman, A.H. "The Role of Science in Defence." *Canadian Aeronautical Journal* 4,9 (September 1958): 213-16.

Index

STUDIES IN CANADIAN MILITARY HISTORY

Canadian War Museum | Musée canadien de la guerre

John Griffith Armstrong, *The Halifax Explosion and the Royal Canadian Navy: Inquiry and Intrigue*

Andrew Richter, *Avoiding Armageddon: Canadian Military Strategy and Nuclear Weapons, 1950-63*

William Johnston, *A War of Patrols: Canadian Army Operations in Korea*

Julian Gwyn, *Frigates and Foremasts: The North American Squadron in Nova Scotia Waters, 1745-1815*

Jeffrey A. Keshen, *Saints, Sinners, and Soldiers: Canada's Second World War*

Desmond Morton, *Fight or Pay: Soldiers' Families in the Great War*

Douglas E. Delaney, *The Soldiers' General: Bert Hoffmeister at War*

Michael Whitby, ed., *Commanding Canadians: The Second World War Diaries of A.F.C. Layard*

Martin Auger, *Prisoners of the Home Front: German POWs and "Enemy Aliens" in Southern Quebec, 1940-46*

Tim Cook, *Clio's Warriors: Canadian Historians and the Writing of the World Wars*

Serge Marc Durflinger, *Fighting from Home: The Second World War in Verdun, Quebec*

Richard O. Mayne, *Betrayed: Scandal, Politics, and Canadian Naval Leadership*

P. Whitney Lackenbauer, *Battle Grounds: The Canadian Military and Aboriginal Lands*

Cynthia Toman, *An Officer and a Lady: Canadian Military Nursing and the Second World War*

Michael Petrou, *Renegades: Canadians in the Spanish Civil War*

Amy J. Shaw, *Crisis of Conscience: Conscientious Objection in Canada during the First World War*

James G. Fergusson, *Canada and Ballistic Missile Defence: Déjà Vu All Over Again*

Serge Marc Durflinger, *Veterans with a Vision: Canada's War Blinded in Peace and War*

Benjamin Isitt, *From Victoria to Vladivostok: Canada's Siberian Expedition, 1917-19*

James Wood, *Militia Myths: Ideas of the Canadian Citizen Soldier, 1896-1921*

Timothy Balzer, *The Information Front: The Canadian Army and News Management during the Second World War*